Topics in Applied Physics Volume 10

Topics in Applied Physics Founded by Helmut K. V. Lotsch

Transient
Electromagnetic Fields

Edited by L. B. Felsen

With Contributions by
C. E. Baum   L. B. Felsen   J. A. Fuller   R. Mittra
D. L. Sengupta   C.-T. Tai   J. R. Wait

With 111 Figures

Springer-Verlag Berlin Heidelberg GmbH 1976

Professor LEOPOLD B. FELSEN

Office of the Dean of Engineering,
and Department of Electrical Engineering and Electrophysics,
Polytechnic Institute of New York, Long Island Center, Route 110
Farmingdale, NY 11735, USA

ISBN 978-3-662-30907-0   ISBN 978-3-540-38095-5 (eBook)
DOI 10.1007/978-3-540-38095-5

Library of Congress Cataloging in Publication Data. Main entry under title: Transient electromagnetic fields. (Topics in applied physics; v. 10). Includes bibliographical references and index. 1. Electromagnetic fields. 2. Transients (Electricity) I. Felsen, Leopold B. II. Baum, Carl E., 1940—. QC665.E4T7 530.1'41 75-38548

© Springer-Verlag Berlin Heidelberg 1976
Originally published by Springer-Verlag Berlin Heidelberg New York in 1976
Softcover reprint of the hardcover 1st edition 1976

Monophoto typesetting and offset printing: Zechnersche Buchdruckerei, Speyer. Brühlsche Universitäts-druckerei, Giessen

Preface

While all wave phenomena are causal and hence time-dependent, electromagnetic sources and generators usually operate in the steady-state time-harmonic regime. Hence, there has been a strong emphasis in electromagnetics on time-harmonic wave propagation. However, in recent years, there has emerged a renewed interest in transient wave phenomena, stimulated by various applications that require the explicit treatment of time-dependent effects. One such application is to signal transmission through time-varying media as exemplified by the ionosphere or some other ionized plasma. Another is due to the present ability to produce short electromagnetic pulses with a broad frequency spectrum, and the consequent development of new time-domain techniques for the transmission, reception and scattering characteristics of antennas and targets. Also, short pulses of high power, especially in the optical frequency range, are finding application as diagnostic tools for the study of ablation, implosion and other wave-material interactions. Finally, the effects of impulsively excited electromagnetic bursts on electronic equipment, devices and installations have become a matter of concern. Such bursts may range from naturally caused lightning discharges to man-made nuclear explosions. The security and reliability of communication channels under the influence of such bursts has motivated extensive investigations by private and public organizations concerned with communications, and also by military agencies.

The purpose of this book is to present an overview of the techniques that are employed in the study of transient electromagnetic wave phenomena. The task is addressed by providing a mix between tutorial and educational background material, and up-to-date applications to problem areas of current interest. The tutorial function is served primarily in Chapter 1, which summarizes the basic properties of transient electromagnetic fields, and applies these to wave propagation and diffraction problems in non-dispersive and dispersive media. It may here be remarked that in the conventional teaching of electromagnetic theory, the subject of wave propagation is usually introduced via the time-harmonic regime. This practice is based on the assumption that time-harmonic wave processes can be described more readily than transient processes, the latter being derivable from the former by the additional complication of the Fourier or Laplace transform. However, the basic phenomena of wave propagation, i. e., of a progressive disturbance that travels from a source to a receiver through an ambient environment, are more easily understandable in the transient state, which permits direct signal tracking. The time-harmonic field then emerges as the special case of a continuously emitted excitation at constant frequency. It may also be remarked that the Fourier or Laplace transform route from the time-harmonic regime does not provide the only approach to transient solutions. Sometimes, a direct analysis of a transient problem is considerably simpler, and even easier than the solution for the time-harmonic case. This aspect is illustrated on various examples throughout this book.

The presentation in Chapter 1 emphasizes separation-of-variables techniques and their application to illustrative "separable" problems. In the more general "non-separable" case, one must employ other procedures. Integral equation methods and their numerical treatment are discussed in Chapter 2 and are applied there to a variety of scattering problems. A recently developed procedure, the singularity expansion method, is presented in Chapter 3 and is illustrated there on a representative collection of examples. Chapters 4 and 5 deal with more specific problem areas. In Chapter 4, integral equation and numerical methods are applied to transient excitation and reception of linear antennas, whereas Chapter 5 treats the effects of dispersion on electromagnetic pulse propagation through a realistically modeled earth.

Whenever possible, an effort has been made to provide a physical interpretation of the calculated fields so that the reader will gain some insight into the various wave processes that are operative under different conditions. It is hoped thereby to render this volume useful to the uninitiated who would like to familiarize himself with transient wave phenomena, and to the specialist who would like to refer to specific examples. The references appended to each chapter are quite extensive so that source material and related contributions to the literature can be consulted. Although the stress here is on electromagnetic fields, the techniques and results are representative also of other wave fields encountered in underwater sound, elastic media, and similar areas. The material should therefore be useful also to workers concerned with transient wave fields in other disciplines.

In a volume that contains contributions by several authors, coordination of style and notation, and the avoidance of duplication, are a major concern. While the editor has sought to achieve at least a minimal degree of uniformity, he cannot claim unqualified success. Notation does change from one chapter to the next, but it has been attempted to have all symbols properly defined, thereby minimizing the possibility of confusion and misunderstanding. The following conventions have, however, been adhered to throughout: When time-harmonic fields of frequency ω are discussed, the time convention is $\exp(-i\omega t)$; the Laplace transform variable is $s = -i\omega$; a bold faced symbol denotes a vector quantity, and a bold faced symbol with a wiggly underline denotes a dyadic quantity.

Finally, the editor would like to express his thanks to the authors who have contributed to this volume: Professor RAJ MITTRA for Chapter 2, Dr. CARL E. BAUM for Chapter 3, Dr. DIPAK L. SENGUPTA and Professor CHEN-TO TAI for Chapter 4, and Dr. JAMES A. FULLER and Professor JAMES R. WAIT for Chapter 5. Their cooperation is responsible for whatever cohesive thread runs throughout this volume. Also acknowledged with appreciation are the coordinating and editorial services of Mrs. ANNA MAE CUOMO of the Polytechnic Institute of New York, and her expert typing and editing of Chapter 1, which was contributed by the writer.

Farmingdale, N.Y.   LEOPOLD B. FELSEN
September 1975

Contents

Contributors

BAUM, CARL E.
Air Force Weapons Laboratory, Kirtland Air Force Base, Albuquerque, NM 87117, USA

FELSEN, LEOPOLD B.
Office of the Dean of Engineering, and Department of Electrical Engineering and Electrophysics, Polytechnic Institute of New York, Long Island Center, Route 110, Farmingdale, NY 11735, USA

FULLER, JAMES A.
1316 Dunbarton Drive, Richardson, TX 75080, USA

MITTRA, RAJ
Electromagnetics Laboratory, Department of Electrical Engineering, University of Illinois, Urbana, IL 61801, USA

SENGUPTA, DIPAK L.
TAI, CHEN-TO
The Radiation Laboratory, Department of Electrical and Computer Engineering, The University of Michigan, Ann Arbor, MI 48104, USA

WAIT, JAMES R.
CIRES, University of Colorado, Boulder, CO 80302, USA

1. Propagation and Diffraction of Transient Fields in Non-Dispersive and Dispersive Media

L. B. FELSEN

With 16 Figures

The presentation in this chapter has a threefold purpose: to review analytical techniques for dealing with various kinds of transient propagation and scattering, to provide physical interpretations that clarify the meaning of the mathematical results, and to illustrate general concepts on specific examples. This task has been undertaken before. However, in a volume on transient electromagnetic fields, a tutorial introduction is not out of place. In addressing the intended goal, it was considered necessary to repeat a good deal of material available elsewhere in order to make the discussions sufficiently self-contained. However, an effort has been made to inject different points of view and to provide a cohesiveness that is not as easily found in a recent volume [1.1] from which the present chapter draws much of its substance. It is hoped that the overview presented here will make it easier for the less specialized reader to follow up other literature sources.

Turning to the specific contents of this chapter, we begin in Section 1.1 with some general observations on transient wave processes and how the different chapters in this book relate to them. Section 1.2 then reviews the electromagnetic equations and boundary conditions, dyadic and scalar Green's functions, and also the distinction between a non-dispersive and dispersive formulation of field problems in a material medium. Sections 1.3 and 1.4 deal with analytical methods for transient fields. Various eigenfunction expansions for separable problems are reviewed in Section 1.3, and Fourier or Laplace integral methods are then singled out in Section 1.4. Included are exact inversions of Laplace integrals, the relation between the transient wavefront and the high-frequency time harmonic regimes, the use of rays for tracking high-frequency fields, and general asymptotic (saddle point) methods for reducing integral representations for time-harmonic fields and for time-dependent fields in a dispersive environment. The general conclusions inferred from Sections 1.3 and 1.4 are then illustrated on various examples dealing with non-dispersive (Sect. 1.5) and dispersive (Sect. 1.6) propagation.

1.1 Introduction

The behavior of transient fields depends strongly on whether or not a propagation channel is dispersive. In a dispersive but stationary environment, the wave propagation speed changes with the wave frequency whereas in a dispersionless stationary environment, all frequencies propagate at the same speed. Since a time-dependent input signal can be resolved into its spectral frequency components, each such component can be tracked through a dispersive propagation medium, and the result combined at the receiver. Depending on the frequency profile of the input signal and the dispersive properties of the medium, the received signal may be smeared out, compressed, or otherwise affected. Dispersive effects are generally negligible at observation times close to the time of arrival of the first response (the wavefront) since synthesis of the transition from a zero field before the arrival to a finite field after passage of the wavefront requires a high-frequency content; at sufficiently high frequencies, the particle constituents of a physical medium cannot respond to the incident excitation and therefore render it dispersion-free. On the other hand, at very long observation times, when low frequency constituents are important, the signal response is governed by the natural or resonant frequencies ω_i of the propagation environment. Between the early-time response near the wavefront and the long-time "ringing" at the natural frequencies, the propagation process may be described in terms of wave packets with small frequency spread about a fixed central value. In a homogeneous medium, a wave packet travels with constant speed and an observer moving at that speed therefore sees himself surrounded by fields oscillating at a constant frequency. A stationary observer is reached by wave packets with increasingly lower speeds and thus measures fields whose frequency changes with time. These aspects are schematized in Fig. 1.1.

When the ambient medium is non-dispersive, but a transient field excites a scattering object of finite size, the phenomenology is somewhat similar. The field near the time of arrival of the incident wavefront is again determined by its high-frequency content. However, at later observation times, long after the wavefront has traversed the scatterer, the induced currents (if the object surface is perfectly conducting) set up oscillations at the natural (resonance) frequencies of the scatterer. In a completely enclosed non-dissipative environment, the natural oscillations are undamped. In an open region, leakage of energy to infinity leads to damped oscillations, and the most weakly damped dominate the long-time response; the field is then well described by oscillations at one or a few preferred frequencies.

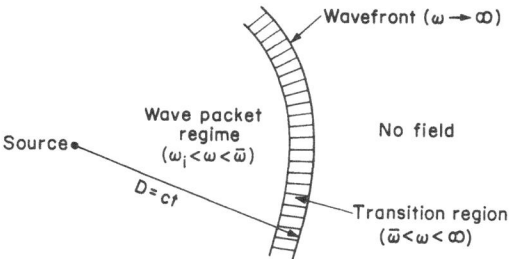

Fig. 1.1 Transient propagation processes. When a source in a dispersive medium is turned on at time $t=0$, the wavefront at time t has moved without dispersion to $D=ct$, where c is the highest propagation speed. In a transition region behind the wavefront, dispersion becomes effective, and outside the transition region, propagation may be described by fully developed dispersive wave packets. ω_i is a natural frequency of the medium, and $\bar{\omega} \gg \omega_i$ is a frequency at which dispersion becomes noticeable. The spectral frequency ranges that contribute to the transient field in various regions have been indicated in the figure

From these observations, it is apparent that the long-time response of a finite scatterer is determined primarily by its overall size rather than by its detailed shape. The opposite is true of the early-time response. The wavefronts are exceedingly sensitive to structural features and therefore provide fine tools for probing the contour of a scattering object. This is illustrated in Fig. 1.2 where a plane wavefront is incident on a triangular cylindrical scatterer. After reaching edge 1, the incident wavefront A sets up a reflected plane wavefront B and a circular cylindri-

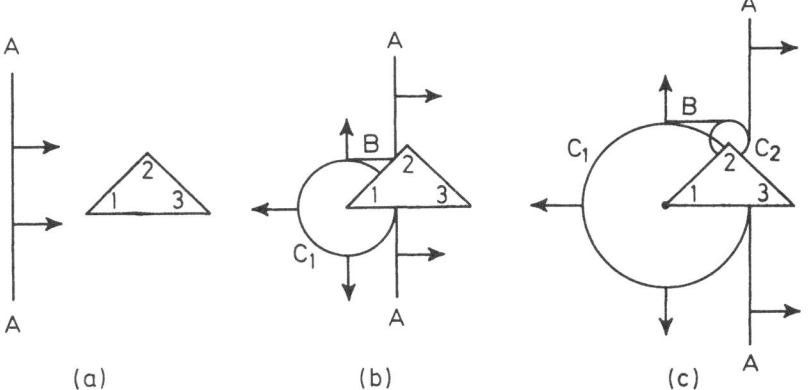

Fig. 1.2a–c. Wavefronts due to scattering by a triangular cylinder. a) Incident wavefront A before reaching scatterer. b) Wavefronts after A reaches edge 1 but before reaching edge 2; $B=$ reflected front originating at edge 1. c) Wavefronts after A reaches edge 2 but before reaching edge 3; $C_2=$ diffracted front originating at edge 2

cal ("diffracted") wavefront C_1, propagating in the direction of the arrows. After wavefront A passes edge 2, it sets up another cylindrical wavefront C_2, and similarly for edge 3. When one of the cylindrical wavefronts reaches another edge, it again sets up a new cylindrical wavefront propagating away from that edge, etc. Since the various wavefronts reach an observer at distinct arrival times (although they all propagate at the same speed c), they evidently provide an intricate pattern of information on the structural details of the scattering object.

In this book, it has been attempted to elucidate in mathematical and physical terms various transient wave processes such as those described above, to provide the mathematical and numerical machinery required for the extraction of specific solutions, and to include a representative set of illustrative examples, most of which are accompanied by plotted numerical results. As noted earlier, the discussion in Chapter 1 concentrates primarily on a review of the properties of transient electromagnetic fields and on eigenfunction expansion methods that can be applied to separable problems, i.e., to configurations for which the space-time partial differential field equations can be reduced to a set of ordinary differential equations in a single variable. Alternative field representations are emphasized, which exhibit various convergence properties and are useful for different ranges of the parameters descriptive of a problem. Examples include pulse diffraction by perfectly conducting and dielectric configurations, and pulse propagation in dispersive media, such as an ionized plasma.

The separation of variables method is inadequate to deal with non-separable antenna or scatter configurations excited by an impinging transient field. In this more general case, the induced surface currents and charges (when the object is perfectly conducting) must be determined by solution of integral equations; after these currents and charges have been calculated, one may evaluate the fields which they generate throughout space. The relevant integral equations are formulated in Chapter 2, with special emphasis on their suitability for treatment by numerical methods. The comprehensive discussion in Chapter 2, and the numerical examples accompanying it, provide an up-to-date exposition of the numerical approach to the solution of time-dependent boundary value problems. This is followed in Chapter 3 by a presentation of the singularity expansion method (SEM), a recently developed and highly successful procedure based on the resonant field configurations that can be supported on bounded metallic objects of relatively arbitrary shape. These natural oscillations, already alluded to earlier, furnish not only an efficient method for calculating the transient response but they also grant some basic insight into the behavior of transient fields as a function of time. A principal motivation for the exploration of SEM came, in fact, from a study of experimental transient response data which exhibited marked regularities interpretable as damped sinusoidal oscillations. This feature is well illustrated by the numerous examples cited in this chapter and elsewhere in the book.

Chapters 4 and 5 deal with specific applications. The techniques of Chapter 2, and also some alternative analytical and numerical schemes, are applied in Chapter 4 to a detailed study of radiation and reception by linear antennas. This presentation elucidates the waveforms generated when a linear antenna operates in the transmitting or receiving mode, and it also demonstrates the waveform control effected by resistive or reactive antenna loading. Again, the theoretical predictions here are accompanied by representative numerical examples.

The emphasis in Chapters 2 to 4 is on transient signal perturbation due to perfectly conducting structural shapes embedded in a non-dispersive medium. However, in a dispersive environment, the signal is distorted even in the absence of structural configurations. This feature has been discussed in general terms in Chapter 1 and illustrated there on some simple idealized examples. A detailed study of the influence of medium dispersion is carried out in Chapter 5 for the problem of signal propagation through the earth, modeled realistically by recourse to experimental data on soil permittivity and conductivity. By use of the analytical methods summarized in Chapter 1, results are obtained for the high and low frequency portions of the response for varying soil conditions and for various depths below the earth's surface.

As the preceding synopsis indicates, the reader will find in this book a rather extensive collection of numerical data on various transient radiation and scattering problems in non-dispersive and dispersive environments, and also an exposition of the various methods, both analytical and numerical, that are currently in use.

1.2 Field Equations and Boundary Conditions

The behavior of the electromagnetic field is governed by the Maxwell equations. In a source-free region, these equations provide relations between the space and time dependent vector electric field $E(r,t)$, the magnetic field $H(r,t)$, the electric flux density $D(r,t)$ and the magnetic flux density $B(r,t)$, with $r=(x,y,z)$ and t denoting the spatial and temporal variables, respectively. The equations also contain constitutive parameters that are descriptive of the medium wherein the electromagnetic field is evaluated. The permittivity ε provides a relation between D and E while the permeability μ provides a relation between B and

H; when the medium varies spatially and (or) temporally, one or both of the constitutive parameters will be a function of r and (or) t. In a non-dispersive, isotropic, homogeneous, time-invariant medium, ε is a factor of proportionality between D and E, and μ is a factor of proportionality between B and H. Isotropy implies that the electric and magnetic properties of the medium are independent of direction; when this is not the case, as in anisotropic media (for example, crystals, ferrites, magnetoplasmas), the scalars ε and (or) μ are replaced by dyadics so that the vector D is not parallel to E and (or) B is not parallel to H. By the non-dispersive property, the value of D at a point r_1 and at time t_1 depends only on E and ε at the same (local) coordinates (r_1, t_1). If the medium is dispersive, the value of D at (r_1, t_1) depends (non-locally) also on the electric field at other points r and (or) at other (earlier) times $t < t_1$. When the dependence is on the electric field at other points r only but not on earlier times, the medium is called spatially dispersive, while a dependence on earlier times $t < t_1$ but not on other spatial locations characterizes a temporally dispersive medium. Under the most general conditions, a medium is spatially and temporally dispersive. In the dispersive regime, ε generally becomes an operator that contains spatial and temporal derivatives and their inverses (i.e., integrations). The relation between D and E then involves the non-local operations of space-time differentiation or integration, as illustrated in an example in Subsection 1.2.2. Analogous considerations apply to the relation between B and H in a dispersive environment. As noted in Section 1.1, the non-dispersive assumption can be used only when the incident electromagnetic field does not excite any of the natural resonance phenomena in a material medium; the persistence of resonant effects gives rise to the non-local behavior noted above. Since the resonances are associated with certain natural frequencies, it follows that the validity of a temporally non-dispersive model is limited to input signals whose spectrum is confined to temporal frequency bands well removed from the resonant frequencies.

In this section, we shall examine the formulation of electromagnetic field problems in non-dispersive and dispersive media. We consider also the reduction of the complexity of the vector field problem by the introduction of dyadic Green's functions and scalar potentials.

1.2.1 Non-Dispersive Media

In a non-dispersive, isotropic medium with space and time varying permittivity $\varepsilon(r,t)$ and permeability $\mu(r,t)$, the Maxwell field equations take the form

$$\nabla \times \boldsymbol{H} = \frac{\partial \boldsymbol{D}}{\partial t} + \boldsymbol{J}, \quad \nabla \times \boldsymbol{E} = -\frac{\partial \boldsymbol{B}}{\partial t} - \boldsymbol{M}, \tag{1.1}$$

where $\boldsymbol{J}$ and $\boldsymbol{M}$ are applied electric and magnetic source current densities, respectively, and all quantities depend on (r,t). The relation between $\boldsymbol{D}$, $\boldsymbol{E}$ and $\boldsymbol{B}$, $\boldsymbol{H}$ is provided by

$$\boldsymbol{D}(r,t) = \varepsilon(r,t)\boldsymbol{E}(r,t), \quad \boldsymbol{B}(r,t) = \mu(r,t)\boldsymbol{H}(r,t). \tag{1.2}$$

Electric and magnetic charge densities $\varrho(r,t)$ and $\varrho_m(r,t)$ are associated with the source currents via

$$\nabla \cdot \boldsymbol{J} = -\frac{\partial \varrho}{\partial t}, \quad \nabla \cdot \boldsymbol{M} = -\frac{\partial \varrho_m}{\partial t} \tag{1.3}$$

whence it follows upon taking the vector divergence of (1.1), with (1.3), that at any instant of time,

$$\nabla \cdot \boldsymbol{D} = \varrho, \quad \nabla \cdot \boldsymbol{B} = \varrho_m. \tag{1.4}$$

To render the specification of the field problem unique, it is necessary to specify initial conditions in the time domain, and boundary conditions across any sudden changes in the spatial or temporal properties of the medium. The initial, or causality, condition is based on the physical requirement that no field response be detected before the onset of source excitation. If the source is initiated at $t = t_1$, this implies that at any observation point r,

$$\boldsymbol{E}, \boldsymbol{H} \equiv 0 \quad \text{for} \quad t < t_1. \tag{1.5}$$

A stationary interface S separating two different media is characterized by an abrupt change in the spatial properties of the constitutive parameters across S. If $\boldsymbol{E}_\alpha$, $\boldsymbol{H}_\alpha$, ε_α, μ_α, $\alpha = 1,2$, denote quantities associated with medium 1 and 2 respectively, and if $\boldsymbol{v}_0$ is the unit normal to S directed into medium 2, then it follows trom the field equations that at any time t,

$$\boldsymbol{v}_0 \times (\boldsymbol{H}_2 - \boldsymbol{H}_1) = \boldsymbol{J}_s, \quad \boldsymbol{v}_0 \times (\boldsymbol{E}_2 - \boldsymbol{E}_1) = -\boldsymbol{M}_s, \tag{1.6a}$$

$$\boldsymbol{v}_0 \cdot (\boldsymbol{B}_2 - \boldsymbol{B}_1) = \varrho_{sm}, \quad \boldsymbol{v}_0 \cdot (\boldsymbol{D}_2 - \boldsymbol{D}_1) = \varrho_s, \tag{1.6b}$$

where $\boldsymbol{J}_s$, $\boldsymbol{M}_s$, ϱ_s and ϱ_{sm} are electric and magnetic surface current and charge distributions on S. Equations (1.6) express the requirement of continuity of the tangential components of E and H, and of the normal components of B and D, across any charge-free and current-free interface; i.e., across any interface between two non-perfectly conducting media unless currents and charges are impressed externally. If one of the media (medium 2) is assumed to be perfectly conducting so that the fields in region 2 vanish identically, the induced electric surface currents and charges at any time t are related to the fields at the surface as follows

$$H_1 \times v_0 = J_s, \qquad v_0 \cdot D_1 = -\varrho_s, \qquad (1.7\,\mathrm{a})$$

and furthermore,

$$v_0 \times E_1 = 0, \qquad v_0 \cdot B_1 = 0. \qquad (1.7\,\mathrm{b})$$

When the surface S possesses edges or corners, the boundary conditions must be augmented by an "edge condition" or "tip condition" that restricts the allowable growth of any field component as such a structural irregularity is approached [1.2]. This restriction is based on the physical requirement that the energy content of the field in any finite spatial volume must be bounded. For an impenetrable surface, the edge condition states that, at any time t_1

no component of E or H can grow more rapidly than $d^{-1+\delta}$,

$\delta > 0$. $\qquad (1.8\,\mathrm{a})$

Near a conical tip,

no component of E or H can grow more rapidly than $d^{-3/2+\delta}$,

$\delta > 0$. $\qquad (1.8\,\mathrm{b})$

Here, d is the distance from the edge or tip, respectively. The conditions in (1.8a) and (1.8b) can be used to check the validity of induced surface current solutions obtained, for example, by numerical techniques.

A temporal discontinuity in the constitutive parameters may be produced if a medium changes suddenly, but uniformly throughout all space, at time t_0. It is then required that [1.3].

D and B are continuous at $t = t_0$. $\qquad (1.9)$

The preceding conditions must be generalized for a moving interface where the discontinuities in the constitutive parameters depend on both r and t.

1.2.2 Dispersive Media

The observations about field descriptions in dispersive media, made in the beginning of this section, are well illustrated by a simple example [1.4]: a plasma, consisting of a neutral mixture of charged particles (electrons and ions) in thermal motion. An externally impressed electromagnetic field imparts order to the particle motion, and the resulting secondary fields affect the overall properties of waves propagating through the medium. If the impinging field has only a frequency content that is well above the natural resonant frequencies of the (heavy) ionic constituents, the (light) electrons alone need be considered as mobile in response to this excitation while the ions provide a stationary neutralizing background. One may then model the plasma as a compressible fluid ("warm" plasma), for which one may define a background electron pressure p_0 and mass density $n_0 m$ (here assumed to be homogeneous and stationary), with n_0 and m denoting the electron density and mass, respectively (in a "cold" plasma, modeled as an incompressible fluid, one sets $p_0 = 0$). An externally applied electromagnetic field causes pressure perturbations $p(r,t)$ and electron velocities $v(r,t)$ that are related by the Euler field equations [1.4]

$$\frac{1}{\gamma p_0} \frac{\partial p}{\partial t} + \nabla \cdot v = 0 \tag{1.10a}$$

$$\nabla p + n_0 m \frac{\partial v}{\partial t} = -n_0 q E , \tag{1.10b}$$

where γ is the specific heat ratio for electrons and q is the electronic charge. It is assumed here that the perturbations about the equilibrium state are so small as to render the linear theory applicable; also, dissipative effects are neglected. (For dispersive effects introduced by dissipation in a dielectric medium, see Chapt. 5). The $(-n_0 q E)$ term in (1.10b) represents the vector force exerted by the impressed electric field on the electrons in the fluid. In turn, the moving electrons constitute a current $J^{(e)} = -n_0 q v$ that must be included in the Maxwell equations (1.1), in addition to the externally applied source currents J and M

$$\nabla \times H = \varepsilon_0 \frac{\partial E}{\partial t} - n_0 q v + J , \qquad \nabla \times E = -\mu_0 \frac{\partial H}{\partial t} - M . \tag{1.11}$$

It has here been recognized that the background medium for the electromagnetic field, when the plasma is absent, is vacuum with permittivity ε_0 and permeability μ_0.

Equations (1.10) and (1.11) constitute a first-order coupled system that must be solved self-consistently for the electromagnetic field constituents E, H and the dynamical field constituents p, v. It is to be noted that the constitutive parameters ε_0, μ_0 and γp_0, $n_0 m$ appearing in the Maxwell and Euler equations, respectively, are *non-dispersive*. Equations with dispersive parameters result on elimination of one set of field constituents from the first-order coupled system. Thus, dispersion in the wave process of one type of field is indicative of the interaction produced with another type of field, the dependence on which is hidden in the dispersive medium parameters. To obtain the dispersive electromagnetic field equations, one may proceed by performing the operation $\partial/\partial t$ on (1.10b) and substituting $\nabla \partial p/\partial t = -\gamma p_0 \nabla \nabla \cdot v$ from (1.10a) whence

$$\underset{\sim}{K} \cdot v = -\frac{q}{m} \frac{\partial E}{\partial t}, \tag{1.12}$$

where $\underset{\sim}{K}$ is the dyadic operator

$$\underset{\sim}{K} = \frac{\partial^2}{\partial t^2} \underset{\sim}{1} - c_{\mathrm{p}}^2 \nabla \nabla, \qquad c_{\mathrm{p}}^2 = \frac{\gamma p_0}{n_0 m}, \tag{1.13}$$

$\underset{\sim}{1}$ is the unit dyadic, and c_{p} is the acoustic propagation speed in the plasma fluid. The inverse operation $\underset{\sim}{K}^{-1}$ yields a formal solution for v, which is substituted into (1.11), with the result

$$\nabla \times H \equiv \underset{\sim}{\varepsilon} * \frac{\partial E}{\partial t} + J, \qquad \nabla \times E = -\mu_0 \frac{\partial H}{\partial t} - M \tag{1.14}$$

with $\underset{\sim}{\varepsilon}$ representing the permittivity operator

$$\underset{\sim}{\varepsilon} \equiv \underset{\sim}{\varepsilon} \left(\nabla, \frac{\partial}{\partial t} \right) = \varepsilon_0 (\underset{\sim}{1} + \omega_{\mathrm{p}}^2 \underset{\sim}{K}^{-1}), \qquad \omega_{\mathrm{p}}^2 = \frac{n_0 q^2}{\varepsilon_0 m}, \tag{1.15}$$

and ω_{p} denoting the "plasma frequency" descriptive of the natural oscillations (resonances) of the electron plasma. Evidently, the dispersive equations (1.14) are of higher order than the non-dispersive equations (1.10) and (1.11), as may be recognized on removal of the inverse operator $\underset{\sim}{K}^{-1}$ by multiplication of the $\nabla \times H$ equation in (1.14) by $\underset{\sim}{K}$. If retained in their present form, the dispersive equations (1.14) are integro-differential equations since the presence of derivative operators in $\underset{\sim}{K}$ implies that of integral operators in $\underset{\sim}{K}^{-1}$; the resulting operation has been denoted symbolically by $*$. Because $\underset{\sim}{K}$ contains both ∇ and $\partial/\partial t$, these integrations are over space and time, and therefore signify that the field at a space-time point (r_2, t_2) depends on the field values at other points r in the medium and at earlier times $t < t_2$. This property characterizes the warm plasma as a spatially and temporally dispersive medium. In the cold plasma approximation, where $c_p \equiv 0$, the medium is only temporally dispersive. The operation in (1.14) may then be written as [1]

$$\underline{\varepsilon} * \frac{\partial E}{\partial t} = \varepsilon_0 \frac{\partial}{\partial t} E(r_2, t_2) + \varepsilon_0 \omega_p^2 \int_{t_1}^{t_2} E(r_2, t') dt', \qquad (1.16)$$

where t_1 is the time of initiation of the electromagnetic sources. The operator inversion in the spatially and temporally dispersive case is carried out most conveniently in a plane wave (Fourier integral) basis which reduces the operations to algebraic ones. This aspect is discussed further in Section 1.3.

One may infer the importance of temporal dispersion by examining harmonic field constituents of the form $E(r, t) = E_0(r) \exp(-i\omega t)$, where ω is the radian frequency and i is the imaginary unit. It then follows from (1.16), on ignoring the lower limit t_1, which is not relevant for the time-harmonic case, that

$$\underline{\varepsilon} * \frac{\partial E}{\partial t} \to \varepsilon(\omega)(-i\omega) E_0(r) e^{-i\omega t}, \qquad \varepsilon(\omega) = \left(1 - \frac{\omega_p^2}{\omega^2}\right) \varepsilon_0. \qquad (1.17)$$

If the source frequency spectrum is confined to a range $\omega > \omega_0$ and if $\omega_0 \gg \omega_p$, then the effects of dispersion are negligible and $\varepsilon(\omega) \to \varepsilon_0$. This observation illustrates the earlier statement that non-dispersive modeling of a medium may be justified when the frequency spectrum of the excitation is well separated from (and, for the plasma, lies well above) the natural resonance frequencies.

The most general form of the Maxwell field equations in a spatially and temporally dispersive medium with spatial and temporal inhomogeneities is given by (1), with D and B expressed symbolically as [1.5,6]

$$D(r,t) = \underline{\varepsilon}\left(\nabla, \frac{\partial}{\partial t}; r, t\right) * E(r,t), \qquad B(r,t) = \underline{\mu}\left(\nabla, \frac{\partial}{\partial t}; r, t\right) * H(r,t) \qquad (1.18)$$

---

[1] The non-local character of the temporally dispersive permittivity operator $\underline{\varepsilon}(\partial/\partial t)$ for the cold plasma is made evident in an alternative manner by employing the series expansion

$$E(r_2, t') = \sum_{n=0}^{\infty} \left[\frac{(t' - t_2)^n}{n!} \frac{\partial^n E(r_2, t')}{\partial t'^n}\right]_{t' = t_2}$$

in the integral of (1.16). In the resulting representation, the flux density D at (r_2, t_2) depends not only on the electric field value at that point but also on the temporal derivatives of the field. The latter feature is indicative of the need for knowing E at times $t < t_2$.

where the dependence of the $\mathbf{\varepsilon}$ and $\mathbf{\mu}$ operators on r and t signifies spatial and temporal variability. As in the special examples in (1.14) and (1.16), space and time integral operations are implied in the formal notation of (1.18). When the medium is stationary (i.e., does not vary in time), $(\partial/\partial t)(\mathbf{\varepsilon}*\mathbf{E})=\mathbf{\varepsilon}*\partial\mathbf{E}/\partial t$ as in (1.14).

In a stationary, homogeneous, unbounded medium, it is possible to seek plane wave solutions with space-time dependence $A\exp(i\mathbf{k}\cdot\mathbf{r}-i\omega t)$, where A is the vector amplitude, $\mathbf{k}$ is the wave vector, ω is the wave frequency and i is the imaginary unit (this is equivalent to taking the spatial and temporal Fourier transforms of (1.18)). When the plane wave field is inserted into the field equations, the derivative operations are reduced to multiplicative operations by the equivalences $\nabla\to i\mathbf{k}$, $\partial/\partial t\to-i\omega$. Accordingly, the dispersive operators $\mathbf{\varepsilon}(\nabla,\partial/\partial t)$ and $\mathbf{\mu}(\nabla,\partial/\partial t)$ become $\mathbf{\varepsilon}(i,\mathbf{k},-i\omega)$ and $\mathbf{\mu}(i\mathbf{k},-i\omega)$; i.e., the effects of spatial and temporal dispersion are manifested by the dependence of the constitutive parameters for plane wave propagation on $\mathbf{k}$ and ω, respectively. It then follows that the wave propagation speed, which is a function of the constitutive parameters, varies with wavenumber and (or) wave frequency in a spatially and (or) temporally dispersive environment. This property serves as an alternative characterization of a dispersive medium.

1.2.3 Dyadic Green's Functions and Scalarization

The solution of the Maxwell equations (1.1), with (1.2) or (1.18), is facilitated when the arbitrarily specified source functions $J(r,t)$ and $M(r,t)$ are regarded as superpositions of impulsively excited vector current elements via the identity

$$f(r,t)=\int dr'\int dt'f(r',t')\delta(r-r')\delta(t-t'),\tag{1.19}$$

where f stands for either J or M, and the integration extends over the spatial extent and the temporal duration of the source function. The Dirac delta function $\delta(u-u')$ vanishes when $u\neq u'$, is infinite when $u=u'$, with a normalization such that

$$\int\delta(u-u')du'=1,\tag{1.20}$$

where the integration interval includes the point $u'=u$. This property establishes the validity of (1.19). In rectangular coordinates, for example, $r=(x,y,z)$, $dr'=dx'\,dy'\,dz'$ and $\delta(r-r')=\delta(x-x')\,\delta(y-y')\,\delta(z-z')$; in curvilinear coordinates $r=(u,v,w)$, the product of one-dimensional delta functions is weighted by appropriate metric coefficients that ensure that $\int dr'\,\delta(r-r')=1$ over any volume including the point $r'=r$.

The excitation function $f(r',t')\,\delta(r-r')\,\delta(t-t')$ represents a vector point source of strength $f(r',t')$, located at r' and being excited impulsively at $t=t'$. A further simplification of the excitation process is possible by separating the vectorial properties of f from the fundamental source function. This can be done by the identity $f=\mathbf{1}\cdot f$, which implies that the required source function, and hence the response to it, are dyadic operators; the field due to a unit-strength vector source is then obtained by dot product multiplication with the source vector direction. When these considerations are applied to the case of electric current excitation (i.e., $M\equiv0$), one is led from (1.1) and (1.2) to the equations[2]

$$\nabla\times\underline{T}_m(r,r';t,t') = \frac{\partial}{\partial t}\left[\varepsilon(r,t)\underline{Z}(r,r';t,t')\right]\mathbf{1}\,\delta(r-r')\delta(t-t') \qquad (1.21\,\text{a})$$

$$\nabla\times\underline{Z}(r,r';t,t') = -\frac{\partial}{\partial t}\left[\mu(r,t)\underline{T}_m(r,r';t,t')\right] \qquad (1.21\,\text{b})$$

where $\underline{Z}$ is called the dyadic electric Green's function and $\underline{T}_m$ is the dyadic magnetic transfer function. On dot product multiplication by $J(r',t')$, one obtains the vector electric and magnetic fields excited by an impulsive vector electric current element at r'

$$\begin{aligned} E_1(r,r';t,t') &= -\underline{Z}(r,r';t,t')\cdot J(r',t') \\ H_1(r,r';t,t') &= -\underline{T}_m(r,r';t,t')\cdot J(r',t') \end{aligned} \qquad (1.22)$$

where the minus signs have been introduced by convention and the subscript 1 signifies that the fields are due to an electric current source. When the medium is stationary so that μ and ε are independent of t, one may reduce (1.21) to the following second order partial differential equation (vector wave equation) for $\underline{Z}$

$$\nabla\times\left[\frac{1}{\mu(r)}\nabla\times\underline{Z}\right]+\varepsilon(r)\frac{\partial^2}{\partial t^2}\underline{Z}=\mathbf{1}\,\delta(r-r')\frac{\partial}{\partial t}\,\delta(t-t'). \qquad (1.23)$$

From the solution (1.23) for $\underline{Z}$, one may then calculate $\underline{T}_m$ via (1.21b).

---

[2] Since the spectrum of an impulsive source function $\delta(t-t')$ contains all frequencies, the non-dispersive model in (1.21) is not consistent with this excitation. However, if the dyadic Green's functions are used subsequently to synthesize fields excited by source functions $f(r,t)$ with a confined frequency spectrum wherein ε and μ are non-dispersive, the problem formulated in (1.21) is meaningful.

In a similar manner, one may define the dyadic magnetic Green's function $\underset{\sim}{Y}(r,r';t,t')$ and the dyadic electric transfer function $\underset{\sim}{T_e}(r,r';t,t')$, which yield on dot product multiplication with $M(r',t')$ the vector magnetic and electric fields, respectively (denoted by subscript 2), excited by an impulsive vector magnetic current element at r'

$$H_2(r,r';t,t') = -\underset{\sim}{Y}(r,r';t,t') \cdot M(r',t')$$
$$E_2(r,r';t,t') = -\underset{\sim}{T_e}(r,r';t,t') \cdot M(r',t').$$
(1.24)

In a stationary medium, $\underset{\sim}{Y}$ is then found to satisfy (1.23) provided that μ and ε are interchanged, and $\underset{\sim}{T_e}$ is determined from a knowledge of $\underset{\sim}{Y}$ by (1.21b) provided that $\underset{\sim}{Z}$, $\underset{\sim}{T_m}$ and μ are replaced by $\underset{\sim}{Y}$, $-\underset{\sim}{T_e}$ and ε, respectively. The initial and boundary conditions satisfied by the dyadic Green's functions are easily interferred from those in [1.4,7].

Knowledge of the dyadic Green's and transfer functions permits calculation of the fields $E(r,t)$ and $H(r,t)$ excited by arbitrary distributions $J(r,t)$ and $M(r,t)$ of electric and magnetic currents. Thus, with reference to (1.19)

$$E(r,t) = \sum_{i=1}^{2} \int dr' \int dt' \, E_i(r,r';t,t'),$$
(1.25)

and similarly for $H(r,t)$.

The dyadic Green's and transfer functions, to be denoted generically by $\underset{\sim}{G}$, satisfy symmetry and reciprocity conditions that may be determined a priori without solving the boundary value problem [1.4]. We cite here only the following properties that pertain to spatially unbounded, homogeneous and (or) non-moving media. The former case implies invariance of the solution under spatial displacement whence the dependence on r and r' occurs as the difference $r-r'$

$$\underset{\sim}{G}(r,r';t,t') = \underset{\sim}{G}(r-r';t,t').$$
(1.26a)

When the medium is time-invariant, one has similarly,

$$\underset{\sim}{G}(r,r';t,t') = \underset{\sim}{G}(r,r';t-t'),$$
(1.26b)

and when the medium is unbounded, homogeneous and time-invariant,

$$\underset{\sim}{G}(r,r';t,t') = \underset{\sim}{G}(r-r';t-t').$$
(1.26c)

Although introduction of the dyadic Green's functions has removed the dependence on the source configuration from the fundamental electromagnetic field problem, the actual solution of vector wave equations as exemplified by (1.23) is generally exceedingly difficult. Progress can be made when the problem is "scalarizable", i.e., when the vector field is deducible from scalar potential functions. Scalarization is possible only for special boundary configurations and medium variations. Included among these are configurations with translational symmetry along a rectilinear direction z (all cross sections in planes perpendicular to z are identical and homogeneous) or along the radical direction r in a spherical coordinate system. The reduction of the vector field problem to scalar field problems is discussed in the literature [1.1,7,8]. One now seeks scalar Green's functions $G(r,r';t,t')$, from which the dyadic Green's functions are ascertained. These scalar Green's functions satisfy scalar wave equations which, in the case of a homogeneous station-ary, non-dispersive medium, and for z-directed source current elements[3], take the form

$$\left(\nabla^2 - \frac{1}{c^2}\frac{\partial^2}{\partial t^2}\right)G(r,r';t,t') = -\delta(r-r')\delta(t-t'),\qquad(1.27)$$

where $c^2 = (\mu\varepsilon)^{-1}$ is the wave propagation speed (see footnote on p. 13). As in (1.5), causality requires that

$$G \equiv 0 \quad\text{for}\quad t < t'.\qquad(1.28)$$

If perfectly conducting z-invariant boundaries S are present (i.e., cylindri-cal surfaces of arbitrary cross section, with axis parallel to z), the boundary conditions are, with v denoting the direction normal to S,

$$G = 0 \quad\text{or}\quad \frac{\partial G}{\partial v} = 0 \quad\text{on S, for all } t,\qquad(1.29)$$

depending on whether the scalar Green's function G is of the electric or magnetic type. If S possesses edges, tips or corners, the scalar Green's function must be so restricted as to furnish fields that satisfy the condi-tions in (1.8a) and (1.8b).

The plasma discussed in Subsection 1.2.2 provides a simple example of a dispersive medium. It may be shown [1.4] that the dyadic Green's

---

[3] For other orientations of the source currents, one requires a more complicated scalar Green's function $\sigma(r,r';t,t')$ that is related to G as follows:

$$-\nabla_t^2\sigma = G, \text{ where } \nabla_t^2 = \nabla^2 - \partial^2/\partial z^2 \qquad[1.4]$$

functions are now derivable from two scalar Green's functions G_e and G_p subject to (1.28) and defined by the equations

$$\left[\nabla^2 - \frac{1}{c_{e,p}^2}\left(\frac{\partial^2}{\partial t^2} + \omega_p^2\right)\right] G_{e,p}(r,r';t,t') = -\delta(r-r')\delta(t-t'), \quad (1.30)$$

where $c_e = (\mu_0\varepsilon_0)^{-1/2}$ and c_p in (1.13) represent the electromagnetic and acoustic propagation speeds in a plasma characterized by the plasma frequency ω_p in (1.15). Dispersive phenomena are here introduced by the presence of the undifferentiated term $(\omega_p/c_{e,p})^2 G_{e,p}$, which is not encountered in the non-dispersive wave equation (1.27). It may again be recalled (see footnote, p. 13) that since the plasma model of Subsection 1.2.2 accounts only for "high-frequency" phenomena, the Green's functions in (1.30) should be used solely to synthesize fields with an appropriately confined frequency spectrum.

1.3 Eigenfunction Expansions

The source-excited field equations or the corresponding equations for the dyadic and scalar Green's functions in Section 1.2 can be solved in closed form only for a few special medium and geometrical configurations[4]. Solutions for a more general class of problems can be generated by the method of eigenfunction expansion whereby a source-excited field is resolved into source-free constituents, the modes or eigenfunctions, which individually satisfy the source-free field equations and boundary conditions. The eigenfunctions can be calculated explicitly for "separable" configurations that permit reduction of the four-dimensional (r,t) wave equations plus initial and boundary conditions to four separate equations involving only a single space or time coordinate. Orthogonality properties exhibited by the eigenfunctions facilitate the determination of the modal amplitudes required for synthesis, by modal superposition, of Green's functions that characterize fields due to localized sources. When a set of eigenfunctions for one configuration is employed to represent the field in another, the description is complicated by mode coupling that arises because the boundary conditions are now not satisfied by each mode individually; yet, such a procedure provides a method for

---

[4] The method of conical flow accommodates problems that are characterized by absence of characteristic dimensions, as in a wedge [1.9].

dealing with non-separable problems for which the eigenfunctions cannot otherwise be determined[5].

Under conditions of separability, space-time eigenfunctions may be specified in different ways, depending on which of the space-time variables is selected as a symmetry coordinate. The choice is usually dictated by the nature of the boundary value problem under consideration or, if alternative choices are possible, by the particular information desired. The latter statement refers to the fact that alternative field representations, while equally valid as mathematical descriptions, generally exhibit different convergence properties with respect to the actual calculation of the field in a given parameter range.

1.3.1 Basis Functions in Four Variables

To illustrate the above remarks, we consider a very simple example: the Green's function for the interior of a perfectly conducting rectangular box of volume abd. Thus, we seek the solution of (1.27), with (1.28), subject to the boundary conditions

$$G = 0 \quad \text{at} \quad x = 0, a; \quad y = 0, b; \quad z = 0, d. \tag{1.31}$$

The one-dimensional eigenvalue problems, which result on applying the method of separation of variables to (1.27), (1.28), and (1.31), provide complete sets of one-dimensional orthonormal eigenfunctions. The completeness relations are expressed succinctly as representations of the delta functions [1.10]

$$\delta(x - x') = \sum_{m=1}^{\infty} \Phi_m^{(1)}(x) \, \Phi_m^{(1)}(x'), \quad 0 \le (x, x') \le a,$$

$$\Phi_m^{(1)}(x) = \sqrt{\frac{2}{a}} \sin q_m^{(1)} x, \quad q_m^{(1)} = \frac{m\pi}{a} \tag{1.32}$$

$$\delta(y - y') = \sum_{n=1}^{\infty} \Phi_n^{(2)}(y) \, \Phi_n^{(2)}(y'), \quad 0 \le (y, y') \le b,$$

$$\Phi_n^{(2)}(y) = \sqrt{\frac{2}{b}} \sin q_n^{(2)} y, \quad q_n^{(2)} = \frac{n\pi}{b} \tag{1.33}$$

---

[5] For the use of integral equation formulations in connection with numerical methods, see Chapters 2 and 3.

$$\delta(z-z') = \sum_{l=1}^{\infty} \Phi_l^{(3)}(z)\Phi_l^{(3)}(z'), \qquad 0\le(z,z')\le d,$$

(1.34)

$$\Phi_l^{(3)}(z)=\left|\sqrt{\frac{2}{d}}\sin q_l^{(3)} z, \qquad q_l^{(3)} = \frac{l\pi}{d}\right.$$

$$\delta(t-t') = \int_{-\infty}^{\infty} \Phi_\omega^{(4)}(t)\Phi_\omega^{*(4)}(t')d\omega, \qquad -\infty <(t,t')< \infty,$$

(1.35)

$$\Phi_\omega^{(4)}(t) = \frac{1}{\sqrt{2\pi}}\,e^{-i\omega t}$$

where the asterisk in (1.35) denotes the complex conjugate. Because the region is bounded, the constituent eigenfunctions $\Phi_m^{(1)}(x)$, $\Phi_n^{(2)}(y)$, and $\Phi_l^{(3)}(z)$ for the spatial domain form a discrete set with eigenvalues $q_m^{(1)}$, $q_n^{(2)}$, and $q_l^{(3)}$. Since the temporal variable ranges over an infinite domain, the temporal eigenfunctions $\Phi_\omega^{(4)}(t)$ form a continuous set, expressed by a Fourier integral, with the eigenvalue ω distributed continuously along the real axis. (For open regions that admit energy leakage to infinity, the eigenfunctions associated with an unbounded space coordinate also form a continuous set; see Subsect. 1.5.1).

The one-dimensional eigenfunctions in (1.32)–(1.35) can be employed for construction of alternative representations of the source-excited field G. First, by employing an eigenfunction expansion in all four variables,

$$G= \int_{-\infty+i\Delta}^{\infty+i\Delta} d\omega \sum_{m=1}^{\infty} \sum_{n=1}^{\infty} \sum_{l=1}^{\infty} A_{mnl\omega} \Phi_{mnl}(r)e^{-i\omega t}(2\pi)^{-1/2},$$

(1.36)

where

$$\Phi_{mnl}(r)= \Phi_m^{(1)}(x)\Phi_n^{(2)}(y)\Phi_l^{(3)}(z)$$

(1.37)

one ascertains on substitution into (1.27) and use of the delta function representations in (1.32)–(1.35) that the modal amplitude coefficients are given by

$$A_{mnl\omega}=(2\pi)^{-1/2}\Phi_{mnl}(r')e^{i\omega t'}\left(q_{mnl}^2 - \frac{\omega^2}{c^2}\right)^{-1}$$

(1.38)

where

$$q_{mnl}=[(q_m^{(1)})^2+(q_n^{(2)})^2+(q_l^{(3)})^2]^{1/2}.$$

(1.39)

To ensure that $G \equiv 0$ for $t < t'$, analytic continuation from real to complex ω has been invoked to move the integration path above the real axis $(\Delta > 0)$ in the complex ω-plane; this avoids the pole singularities at

$$\omega_{mnl}^{\pm} = \pm c\, q_{mnl} \,. \tag{1.40}$$

Since the integrand decays exponentially in the singularity-free upper half plane $\text{Im}\{\omega\} > 0$ when $t < t'$, it follows on addition of a path segment along the infinite semi-circle and application of Cauchy's theorem to the resulting closed contour that G in (1.36) vanishes as required. One also observes that G depends on t and t' via the difference $t - t'$ as required for a time-invariant medium, see (1.26b).

It should be noted that the product of the four one-dimensional eigenfunctions in (1.36) does *not* form a solution of the source-free wave equation and therefore does not provide an eigenfunction for the problem (although the boundary conditions are satisfied). To obtain an eigenfunction, it is necessary to impose on the four as yet unrelated eigenvalues a condition, the dispersion relation

$$q_{mnl}^2 - \frac{\omega^2}{c^2} = 0 \,, \tag{1.41}$$

as is easily ascertained from an examination of the source-free form of (1.27). It is then evident that the modal amplitudes $A_{mnl\omega}$ diverge when the frequency ω approaches one of the resonant values $\omega_{mnl}^{\pm}$, as is to be expected when a mode in a lossless environment is excited by a source oscillating at its resonance frequency. The determination of eigenfunctions by searching for the singularities in the response to an excitation that sweeps over all frequencies has actually been incorporated into a powerful mathematical procedure for establishing the completeness of modal sets [1.10].

1.3.2 Basis Functions in Three Variables (Oscillatory Modes)

Instead of an expansion in terms of the four one-dimensional eigenfunctions, one may employ expansions in terms of the eigenfunctions for any three of the four variables. The wave equation (1.27) is then not completely algebraized as in (1.38) but is instead reduced to an ordinary differential equation in the remaining variable. If the representation is in terms of the spatial eigenfunctions, one has

$$G = \sum_{m=1}^{\infty} \sum_{n=1}^{\infty} \sum_{l=1}^{\infty} g_{mnl}^{(4)}(t)\, \Phi_{mnl}(\mathbf{r}) \,, \tag{1.42}$$

where the time-dependent amplitude function is defined by [1.11]

$$\left(\frac{1}{c^2}\frac{d^2}{dt^2} + q_{mnl}^2\right) g_{mnl}^{(4)}(t) = \Phi_{mnl}(\mathbf{r}')\delta(t-t'), \tag{1.43}$$

subject to $g_{mnl}^{(4)} \equiv 0$ for $t < t'$. The solution for $t > t'$ is

$$g_{mnl}^{(4)}(t) = -c^2\Phi_{mnl}(\mathbf{r}')\left\{\frac{\exp[-i\omega_{mnl}^+(t-t')]}{2i\omega_{mnl}^+} + \frac{\exp[-i\omega_{mnl}^-(t-t')]}{2i\omega_{mnl}^-}\right\} \tag{1.44}$$

and involves the resonant frequencies defined in (1.40). This result may also be deduced from (1.36) on eliminating the ω integral by residue evaluation at the pole singularities $\omega_{mnl}^\pm$. One observes from (1.42) with (1.44) that the field is represented in terms of eigenmodes with fixed spatial distributions, oscillating undamped in time at frequencies $\pm cq_{mnl}$. When losses are present in the medium, the spatial variation of an eigenmode is unaffected but the oscillations decay with time, thereby implying that $\text{Im}\{\omega_\alpha\} < 0$, where $\alpha = (mnl)$.

When the above procedure is employed for open regions, the oscillations are also damped, due to radiation losses. In open regions, the spatial variation of an eigenmode field as $r \to \infty$ is that of an outgoing spherical wave whose dependence is described by $(1/r)\exp(ik_\alpha r - i\omega_\alpha t)$, where k_α is the wavenumber for the α mode. Since the dispersion relation for a traveling spherical wave in a non-dispersive medium is $k_\alpha c = \omega_\alpha$, it follows that the damped oscillations in time give rise to field amplitudes that grow exponentially with distance. It is therefore to be expected that field representations in terms of discrete natural modes with damped oscillations may be inadequate or even invalid unless $t \gg r/c$. Nevertheless, such representations, which arise in the context of the singularity expansion method described in Chapter 3, are highly useful because the natural resonances of a finite structure play an important part in characterizing its transient behavior. Loosely speaking, the oscillatory modes often suffice to describe the long-time response of a finite scattering object subjected to transient excitation; at very short observation times after the onset of excitation, the natural oscillations have not yet become established and can therefore not determine the field behavior. The distinguishing feature for fields in open regions is the presence, in general, of a continuous spectrum of spatial eigenfunctions, which can only approximately, and under special conditions, be expressed in terms of the discrete natural modes with damped oscillation. A penetrating discussion of this problem may be found in [1.12].

1.3.3 Basis Functions in Three Variables (Guided Modes)

Alternative to a representation in terms of eigenfunctions in the three space coordinates as in (1.42) are representations that utilize two space coordinates and time. In more general configurations, the three possibilities of selecting two out of three space coordinates may lead to fundamentally different field expansions (see Subsect. 1.5.1); in the present case, the choice of one, say x, y, t, is typical of what occurs for the others. Thus, we let

$$G = \sum_{m=1}^{\infty} \sum_{n=1}^{\infty} \int_{-\infty+i\Delta}^{\infty+i\Delta} d\omega \, g_{mn\omega}^{(3)}(z) \, \Phi_m^{(1)}(x) \, \Phi_n^{(2)}(y) e^{-i\omega t} (2\pi)^{-1/2} \qquad (1.45)$$

to find that $g_{mn\omega}^{(3)}$ is defined by

$$\left(\frac{d^2}{dz^2} + \kappa_{mn\omega}^2\right) g_{mn\omega}^{(3)}(z) = -\Phi_m^{(1)}(x') \, \Phi_n^{(2)}(y') e^{i\omega t'} (2\pi)^{-1/2} \delta(z-z') \, ,$$
$$(1.46)$$

where $\kappa_{mn\omega} = [\omega^2/c^2 - (q_m^{(1)})^2 - (q_n^{(2)})^2]^{1/2}$. In addition, $g_{mn\omega}^{(3)} = 0$ at $z = 0, d$. The solution is [1.10]

$$g_{mn\omega}^{(3)}(z) = \frac{\sin(\kappa_{mn\omega} z_<) \sin[\kappa_{mn\omega}(d-z_>)]}{\kappa_{mn\omega} \sin(\kappa_{mn\omega} d)} \Phi_m^{(1)}(x') \, \Phi_n^{(2)}(y') e^{i\omega t'} (2\pi)^{-1/2} \, ,$$
$$(1.47)$$

where $z_>$ and $z_<$ denote the greater or lesser of z and z'. The field is now expressed in terms of eigenfunctions with specified periodicities along the x, y directions (guided modes along z); the periodicity $\kappa_{mn\omega}$ along z is a function of the x, y periodicities and the frequency ω. For given real ω (away from the resonances $\kappa_{nm\omega} d = l\pi$), $\kappa_{mn\omega}$ becomes imaginary when the x, y periodicities are sufficiently large, and the modal amplitude function $g_{mn\omega}^{(3)}$ then decays away from the source plane $z = z'$. Although $\kappa_{mn\omega}$ is a multivalued function in the complex ω-plane, $g_{mn\omega}^{(3)}$ is an even function of $\kappa_{mn\omega}$ and hence single-valued; its only singularities are poles at the zeros of $\sin(\kappa_{mn\omega} d)$, i.e., at the resonant frequencies in (1.40). By residue evaluation, one may convert (1.45) into (1.42).

1.3.4 Generalizations

The preceding simple example demonstrates the choices available for constructing Green's function solutions by eigenfunction expansion.

Under more general, yet separable, conditions where the medium properties are allowed to vary appropriately in space and (or) in time, the choice may be less arbitrary than above. Any departure from the homogeneous, stationary regime requires in (1.36) a knowledge of the correspondingly more complicated eigenfunction spectra. This is not the case for the other representations. Thus, (1.42) remains unchanged when the medium is spatially homogeneous but changes in time. The temporal variation now leads instead of (1.43) to a modified differential equation for the modal amplitudes. Alternatively, when the medium is stationary but stratified along the z coordinate, (1.45) remains intact but the differential equation (1.46) is modified.

The basic ideas introduced above for eigenfunction expansions of scalar Green's functions can be extended to vector fields. Properties of vector eigenfunctions, their use for the representation of vector fields and their possible derivation from scalar eigenfunctions have been studied extensively in the electromagnetic literature. Some applications for nonseparable configurations are given in Chapter 3.

1.4 Fourier or Laplace Inversion of Time-Harmonic Field

1.4.1 Time-Dependent and Time-Harmonic Fields

The most frequently used method for dealing with transient phenomena in time-invariant media is the Fourier or Laplace integral transform technique. Since the time variable in the field equations appears then solely through the derivative operator $(\partial/\partial t)$, a representation in terms of exponential eigenfunctions, as in the Fourier or Laplace transform, removes the temporal dependence. The resulting "time-harmonic" field is of interest in its own right, and from its solution, the transient field is recovered by application of the inverse transform. This procedure has already been employed in Section 1.3 by introduction of the Fourier integral basis (1.35) for expressing the time-dependent Green's function. However, in contrast to Section 1.3, where the solution to an entire field problem has been constructed by the method of separation of variables, the reduction of the time-dependent to the time-harmonic field involves separation of the temporal variable only.

A general Green's function problem, see (1.27)

$$L\left(\nabla, \frac{\partial}{\partial t}; r\right) G(r, r'; \tau) = -\delta(r - r')\delta(\tau), \qquad \tau = t - t', \tag{1.48}$$

subject to the causality condition (1.28), and to spatial boundary conditions, is reduced in the Fourier integral basis (1.35) to

$$L(\nabla, -i\omega; r)G_\omega(r, r') = -\delta(r - r'),\tag{1.49}$$

subject to the same spatial boundary conditions. Here, L is a differential operator containing the spatial and temporal derivatives ∇ and $\partial/\partial t$, and also a dependence on the position coordinate r to allow for medium inhomogeneity. The medium may be dispersive and lossy. From a knowledge of the time-harmonic field $G_\omega(r, r')$ (corresponding to a time dependence $\exp(-i\omega t)$), one obtains the time-dependent field as

$$G(r, r'; \tau) = \frac{1}{2\pi} \int_{-\infty + i\Delta}^{\infty + i\Delta} G_\omega(r, r') e^{-i\omega\tau} d\omega,\tag{1.50}$$

with the positive parameter Δ chosen large enough to ensure that the integration path in the complex ω-plane passes above all of the singularities of G_ω. As noted in Section 1.3, the causality condition is thereby satisfied. The completeness and orthogonality statement in (1.35) implies also the inverse relation to (1.50)[6]

$$G_\omega(r, r') = \int_0^\infty G(r, r'; \tau) e^{i\omega\tau} d\tau, \quad \text{Im}\{\omega\} > 0,\tag{1.51}$$

where it has been noted that $G \equiv 0$ for $\tau < 0$. Since the ω variable is defined in the complex plane, the complex Fourier integral procedure is essentially equivalent to the Laplace transform procedure, which is usually performed in the s variable

$$s = -i\omega, \quad \text{Re}\{s\} > 0.\tag{1.52}$$

If the field is excited by a point source with temporal dependence $h(t)$ for $t > 0$ ($h \equiv 0$ for $t < 0$), then the resulting G is obtained from (1.50) on multiplying by $h(t')$ and integrating over t', see (1.25)

$$G(r, r'; t) = \frac{1}{2\pi} \int_{-\infty + i\Delta}^{\infty + i\Delta} G_\omega(r, r') h(\omega) e^{-i\omega t} d\omega, \quad t > 0,\tag{1.53}$$

---

[6] Note from (1.35) that $\delta(\omega - \omega') = (1/2\pi) \int_{-\infty}^{\infty} e^{i(\omega - \omega')t} dt$, which permits representation of an ω-dependent function G_ω on multiplication by $G_{\omega'}$ and integration over ω'.

where

$$h(\omega) = \int\limits_{0}^{\infty} h(t') e^{i\omega t'} dt'. \tag{1.54}$$

Thus, the time-harmonic field G_ω need not be known for all values of ω but only for those frequencies that are included in the source spectrum $h(\omega)$. The time-harmonic problem may therefore be analyzed by approximate methods applicable in the required frequency range. For physical media with complicated dispersive behavior, the constitutive parameters are often ascertained by measurements performed at various frequencies, and an analytical model is then constructed to accommodate the observed data. The transient response for signals with spectral content inside the accessible range can then be ascertained analytically (see Chapt. 5).

1.4.2 Explicit Inversion

While the evaluation of G from G_ω in (1.50) must generally be done by approximate analytical or by numerical techniques, there exist special cases for which the Fourier inversion can be performed trivially and yields a transient field solution that is often simpler than the time-harmonic solution. Included among these is the class of problems whose time-harmonic field G_ω can be expressed as the Laplace integral (1.51), from which G is then extracted by inspection. The result is stated as follows [1.13]:

Suppose that a time-harmonic Green's function $G_\omega(r, r')$, with a suppressed time dependence $\exp(-i\omega t)$, can be expressed as the contour integral

$$G_\omega(r, r') = \int\limits_{C} Q(r, r'; w) \exp\left[i \frac{\omega}{c} \gamma(r, r'; w)\right] dw \tag{1.55}$$

where C is the integration path in Fig. 1.3 (symmetrical about $w = 0$), c is a constant, and Q and γ are functions independent of ω. Then the time-dependent Green's function is

$$G(r, r'; \tau) = 0, \qquad\qquad \tau < \frac{D}{c}, \quad D = \gamma(r, r'; 0)$$

$$= -2c \frac{\operatorname{Re}\{i Q(r, r'; -i\beta)\}}{(d/d\beta)\gamma(r, r'; -i\beta)}, \quad \tau > \frac{D}{c} \tag{1.56}$$

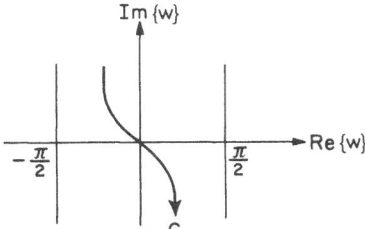

Fig. 1.3. Integration path in complex w-plane

where $\beta = \beta(r, r'; \tau)$ is defined implicitly by

$$c\tau = \gamma(r, r'; -i\beta) \tag{1.57}$$

and it has been assumed that $iQ(r, r'; w)$ is real for real w, and that $\gamma(r, r'; w)$ is an even function of w and is real for real w. Since c usually represents a propagation speed and D a distance, the first of (1.56) expresses causality at an observation point at a finite distance from the source.

Instead of the integral in (1.55), one also encounters time-harmonic Green's functions of the form

$$G_\omega(r, r') = i \int_C H_0^{(1)} \left[\frac{\omega}{c} \gamma(r, r'; w) \right] Q(r, r'; w) \, dw \tag{1.58}$$

where γ and Q are assumed to behave as noted above. By the procedure described in [1.13], this integral may also be converted into a Laplace integral, with the resulting solution for the time-dependent Green's function

$$G(r, r'; \tau) = 0, \qquad \qquad \tau < \frac{D}{c},$$

$$D = \gamma(r, r'; 0) \tag{1.59}$$

$$= -\frac{4}{\pi} \int_0^\beta \frac{\mathrm{Re}\{iQ(r, r'; -i\zeta)\} \, d\zeta}{[\tau^2 - c^{-2}\gamma^2(r, r'; -i\zeta)]^{1/2}}, \qquad \tau > \frac{D}{c}$$

where β is defined implicitly as in (1.57).

A number of propagation and diffraction problems belong to the categories specified above. They include plane, cylindrical and spherical pulse propagation in free space; diffraction by a perfectly conducting or perfectly absorbing wedge; diffraction by a perfectly conducting cone;

transient excitation of a lossless dielectric half space; transient excitation of a unidirectionally conducting infinite or semi-infinite screen. The independence of Q and γ on ω identifies the problems transformable into (1.55) or (1.58) as non-dispersive. Nevertheless, even when Q and β are functions of ω, expansion of these functions into series of powers of ω or $(1/\omega)$ may provide an opportunity for employing the inversion procedure leading to (1.56) or (1.59) on each term of the series.

1.4.3 High-Frequency Time-Harmonic Fields and Transient Fields Near a Wavefront

When the frequency ω tends to infinity[7], the description of time-harmonic fields simplifies substantially. In an isotropic, lossless medium, propagation and diffraction phenomena are then characterized approximately in terms of "local plane waves"

$$\bar{u}_\omega(r) = A(r) \exp\left[i(\omega/c_0)S(r)\right], \tag{1.60}$$

where $A(r)$ is a spatially dependent amplitude, c_0 is a reference propagation speed, and $S(r)$ is the phase or "eikonal". (In a true plane wave, A would be constant and the surface $S(r) = \text{constant}$ would be planar.) The local plane wave description applies not only to an incident field but also to reflected, refracted and diffracted fields. The incident, reflected and refracted fields obey the rules of geometrical optics. Diffracted fields account for penetration into shadow regions and for other phenomena not included in geometrical optics. Their excitation by an incident field is deduced from another set of rules but, after being launched, diffracted fields also propagate according to geometrical-optics laws. Thus, the high-frequency time-harmonic field at an observation point may be synthesized by the sum of all local plane wave fields (incident, reflected, refracted and diffracted) passing through that point. The local plane wave fields propagate along trajectories called "rays", which are perpendicular to the equiphase surfaces $S(r) = \text{constant}$. The rays satisfy Fermat's principle of least propagation time in its original or extended form (to account for diffracted rays) [1.14]. After the family of rays

---

[7] While $\omega \to \infty$ provides a mathematical condition for the validity of high-frequency asymptotic methods, a practical criterion is $kd \gg 1$, where $k \approx \omega/c$ is the wavenumber and d is a characteristic dimension of the propagation or scattering configuration. Thus, the "high-frequency" regime obtains when d is large compared to the local wavelength of the incident field.

for a particular wave species has been determined (this depends on initial conditions in source regions, at reflecting and refracting surfaces, and at launch points of diffracted rays), the eikonal S is found by integrating the refractive index $n(r) = c'(r)^{-1}$, where c' is a normalized propagation speed, along a ray (this "phase integral" furnishes the optical length),

$$S(r) - S(r_1) = \int_{r_1}^{r} n(r) ds ; \tag{1.61}$$

the amplitude A is then determined from considerations of energy conservation in a tube of rays (see Fig. 1.4). In a homogeneous medium

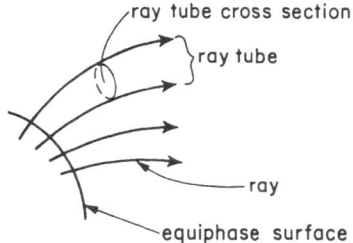

Fig. 1.4. Rays, ray tubes and equiphase surfaces. The spatially dependent amplitude $A(r)$ of the local plane wave field propagating along a ray varies inversely with the square root of the ray tube cross section

with $n =$ constant, the rays are straight lines and (1.61) reduces to $S - S_1 = nD$, where D is the distance along a ray, and S_1 is a phase reference. The above-sketched procedure is known as the geometrical theory of diffraction and has been well documented [1.14,15][8]. The description (1.60) fails in "transition regions" (for example, near light-shadow boundaries) where the field varies so rapidly as not to be characterizable in terms of the much more regular local plane waves; however, the theory can then be "patched up" by use of appropriate transition functions that provide the correct behavior in and near the transition regions.

The local plane wave field (1.60) represents the first term in a formal asymptotic expansion in inverse powers of ω

$$u_\omega(r) \sim \exp\left[i\omega S(r)/c_0\right]\left[A(r) + (-i\omega)^{-\alpha-1} \sum_{m=0}^{\infty} \frac{u_m(r)}{(-i\omega)^m}\right] \tag{1.62}$$

---

[8] For a recent treatment see volume 3 of the same series [1.40].

which has been generalized here by the inclusion of the factor $(-i\omega)^{-\alpha-1}$, where $\alpha > -1$[9]. When this expansion is continued analytically into the complex frequency domain $\text{Im}\{\omega\} > 0$, it follows from (1.51) that the principal contribution to the integral comes from the lower integration limit since $\exp(i\omega\tau)$ decays rapidly with increasing τ. With reference to (1.51), the lower integration limit should actually be $\tau_0 > 0$ since $G(r,r';\tau)$, or in the present case $u(r,\tau)$, at a given observation point vanishes identically before the arrival of the initial response (wavefront) that propagates outward from the source region with finite speed. Thus, the high-frequency behavior of $u_\omega(r)$ is determined via (1.51) by the behavior of $u(r,\tau)$ near $\tau = \tau_0$, and vice versa. Assuming that $u(r,\tau)$ has near τ_0 the series expansion

$$u(r,\tau) \sim \overline{v}(r)\delta(\tau-\tau_0) + (\tau-\tau_0)^\nu \sum_{m=0}^{\infty} v_m(r)(\tau-\tau_0)^m, \qquad \tau > \tau_0 \qquad (1.63)$$

then by insertion into (1.51) (with the lower limit replaced by τ_0), termwise integration, and use of the formula

$$\Gamma(x) = \int_0^\infty y^{x-1}e^{-y}dy, \qquad \text{Re}\{x\} > 0, \qquad (1.64)$$

for the gamma function, one finds that $u_\omega(r)$ is given by (1.62) provided that the following relations are satisfied

$$c\tau_0(r) = S(r), \qquad \overline{v}(r) = A(r), \qquad \nu = \alpha,$$
$$u_m(r) = v_m(r)\Gamma(\nu+m+1). \qquad (1.65)$$

Thus, from a knowledge of the high-frequency asymptotic expansion of the incident, reflected, refracted, or diffracted field, any one of which is here denoted by u_ω, one may infer without further calculation the behavior of the corresponding time-dependent field u near its wavefront defined by the surface $S(r) = c\tau_0(r)$. Since wavefronts belonging to different wave species generally arrive at a given observation point at different times τ_0, the incident, reflected, etc., transient fields are readily resolved (see Fig. 1.2), in contrast to the time-harmonic fields which are distinguished only by different phase relations. One observes that the nature

---

[9] Note that ω-dependent factors occur also when the source function is not an impulse; the time-harmonic Green's function is then multiplied by the source spectrum function $h(\omega)$, see (1.53). If $h(\omega)$ decays at $\omega \to \infty$, this weakens the field discontinuity across the wavefront as described in (1.63).

of the field discontinuity across a wavefront (note that $u \equiv 0$ for $\tau < \tau_0$) is directly related to the time-harmonic field amplitude, with the weaker discontinuities (higher powers of $(\tau - \tau_0)$) corresponding to weaker time-harmonic fields (larger inverse powers of ω).

In the preceding discussion, it has been assumed that a physical medium can, at sufficiently high frequencies, be represented by a frequency independent refractive index, which may be spatially inhomogeneous. As noted earlier, this assumption can be justified only over a restricted range of frequencies that is well removed from the natural resonances of the medium. For improvement, in a homogeneous medium, one may approximate $S = n(\omega)D$ in the high-frequency regime as follows

$$S \sim \left(1 - \frac{\omega_1^2}{\omega^2}\right) D, \tag{1.66}$$

where ω_1 is proportional to a natural resonance frequency of the medium ($\varepsilon(\omega)/\varepsilon_0 = n^2$ in (1.17) provides an example) and a reference phase $S_1 = 0$ has been chosen. The presence of the second term in (1.66) accounts for dispersion. When a typical term from the asymptotic expansion of u_ω in (1.62) is substituted into (1.50), one obtains

$$u_m(r, \tau) \sim \frac{1}{2\pi} u_m(r) \int_{-\infty + i\Delta}^{\infty + i\Delta} \frac{\exp\left[i(\omega/c_0 - \omega_1^2/c_0\,\omega)D - i\omega\tau\right]}{(-i\omega)^\xi} d\omega \tag{1.67}$$

where $\xi = \alpha + m + 1$. The integral in (1.67) may be evaluated in terms of a Bessel function to yield [1.16]

$$u_m(r, \tau) \sim u_m(r) \left(\frac{\bar{\tau}}{\zeta}\right)^{(\xi - 1)/2} J_{\xi - 1}\left[2(\zeta\bar{\tau})^{1/2}\right], \tag{1.68}$$

$$\zeta = \omega_1^2 \tau_0, \quad \bar{\tau} = \tau - \tau_0, \quad \tau_0 = D/c_0.$$

Note that ζ is large since ω_1 is large to be effective in (1.66) and D is finite. For small values of $(\zeta\bar{\tau})$, i.e., very small $\bar{\tau}$, the small argument approximation $J_\nu(x) \sim (x/2)^\nu / \Gamma(\nu + 1)$ reduces (1.68) to

$$u_m(r, \tau) \sim u_m(r) \bar{\tau}^{\xi - 1} / \Gamma(\xi), \tag{1.69}$$

in agreement with (1.63) and (1.65). Thus, very near the wavefront, the relevant frequency constituents are so high that dispersion is not effective. Dispersion becomes noticeable when $2(\zeta\bar{\tau})^{1/2}$ assumes moderately large values, and the field behaves like a wave packet (see Subsects. 1.4.4 and 1.6) when the large argument approximation,

$$J_\nu(x) \sim (2/\pi x)^{1/2} \cos(x - \nu\pi/2 - \pi/4), \qquad x \gg \nu, \tag{1.70}$$

becomes applicable ($\zeta\bar\tau$ large but $\bar\tau$ still small):

$$u_m(r, \tau) \sim u_m(r) \frac{1}{\sqrt\pi} \frac{\bar\tau^{\xi/2 - 3/4}}{\zeta^{\xi/2 - 1/4}} \cos\left[2(\zeta\bar\tau)^{1/2} - (\xi - 1)\frac{\pi}{2} - \frac{\pi}{4} \right]. \tag{1.71}$$

Thus, (1.68) furnishes a transition function that permits the field in a dispersive medium to be evaluated very near the wavefront, where dispersion is negligible because the field is synthesized by very high frequency constituents, and also sufficiently far behind the wavefront where dispersion of somewhat lower frequency constituents leads to the formation of wave packets. The wave packet frequency $\bar\omega$ is ascertained from the time derivative of the argument of the cosine function

$$\bar\omega = \omega_1(\tau_0/\bar\tau)^{1/2}, \tag{1.72}$$

and remains constant if the observer moves at constant speed (i.e., $\tau_0/\bar\tau = $ constant). The corresponding wavenumber $k(\bar\omega) = S(\bar\omega)\bar\omega/c_0 D$ then follows from (1.66) or from the space derivative of the argument of the cosine function. For the case of a cold plasma, these expressions agree with those in (1.140) and (1.141) as $\bar\tau \to 0$.

In an inhomogeneous medium with refractive index $n = n(r, \omega)$, S is given by the phase integral (1.61), and (1.66) is therefore replaced by

$$S \sim \int_{r_1}^{r} [1 - \omega_1^2(r)/\omega^2] ds. \tag{1.73}$$

The medium inhomogeneity is here expressed by the variable resonance frequency parameter $\omega_1(r)$ which in a plasma, for example, is indicative of a variable electron density. The integral in (1.73) is again taken along the ray path (since the refractive index deviates only slightly from unity in the high-frequency range under consideration, the ray path may still be taken as the straight line between r_1 and r). When (1.73) is employed in (1.67), the only modification arises from the replacement of

$$\omega_1^2 D \quad \text{by} \quad \int_{r_1}^{r} \omega_1^2(r) ds.$$

When ζ in (1.68) is modified accordingly, the expressions in (1.68), (1.69), and (1.71) apply also to the transition region behind the wavefront in an inhomogeneous medium.

1.4.4 Contour Integral Methods

When the Fourier inversion of the time-harmonic field in (1.50) cannot be performed by the method of Subsection 1.4.2, one may resort to contour integral techniques in the complex ω plane for simplification of the integral representation. G_ω possesses singularities in the region below the integration path. By contour deformation about these singularities, which may be in the form of poles and branch points, one may express G in (1.50) in terms of residue contributions from the poles and branch cut integral contributions from the branch points. Such a path deformation into the lower half of the ω plane, and the required vanishing of the integral over an infinite semicircle, is made possible by the decay of $\exp(-i\omega\tau)$ when $\tau > 0$. The evaluation of the residue fields at poles ω_α requires only a knowledge of G_ω at ω_α. Contributions from branch points involve an integration of G_ω over a suitable contour surrounding each branch point and the associated branch cuts (branch cuts are introduced to render the definition of G_ω unique on a multi-sheeted ω plane). Depending on the representation employed for G_ω, it may be possible to evaluate the branch cut integrals analytically. Alternatively, it may be possible to employ numerical integration. As already mentioned, G_ω is generally not known in closed form but only in the form of an eigenfunction expansion, which may involve infinite series[10] and (or) integrals. Thus, even when the ω integration in (1.50) can be carried out, the remaining series or integrals remain to be dealt with.

1.4.5 Asymptotic Methods

Although the availability of high-speed computers has made it possible, in many instances, to generate accurate field values by numerical evaluation of the series or integrals appearing in the field representations, it is often adequate to employ simpler approximate analytical formulas. Even when numerical methods are preferred for greater accuracy, approximate analytical results are valuable because they can facilitate an under-

---

[10] Note that series expansions can be expressed in integral form whence integral representations constitute the most general formulation of a field problem [1.10].

standing of the numerical data and delineate the dependence of the solution on various parameters occurring in the field problem. Asymptotic methods for evaluation of integrals afford a powerful means for obtaining approximate analytical solutions with strong physical content. By these techniques, which require some parameter of the problem to be large, the value of an integral is approximated by contributions from critical points in the complex integration plane. These critical points are stationary (saddle) points, poles and branch points of the integrand, and endpoints of the integration interval. The procedure in general is described, for example, in [1.17], and for the inversion of the Fourier integral (1.50), in particular, in [1.18].

Each contributing critical point characterizes a fundamental propagation or scattering process. The association of a particular wave process with a specific type of critical point depends on the choice of representation for the Green's function; for example, in one representation, the phenomenon may be associated with a stationary point and in another, with a pole. Since the integration variables usually denote physical quantities (such as frequency or wavenumber), results derived from alternative representations provide different insights into a particular propagation or scattering mechanism. In the time-harmonic regime, the asymptotic solutions are of the high-frequency type and lead to a ray-optical interpretation of the field; contributions from various critical points then yield the species of incident, reflected, refracted, and diffracted ray fields discussed in Subsection 1.4.3. In the time-dependent regime, in dispersive media, the asymptotic results describe wave packets.

The behavior of the high-frequency, time-harmonic field has already been given in (1.60) and (1.62), and its construction has been outlined. When these results are deduced from integral representations derived by separation of variables, they typically require the evaluation of integrals of the form

$$I_\omega = \int_P V(r, r'; \xi) \exp[ik\psi(r, r'; \xi)] d\xi, \quad k = \omega/c, \quad (1.74)$$

where r is the observation point coordinate, r' is the source coordinate, and the integration variable ξ plays the role of a wavenumber along one of the space coordinates. The phase function ψ is exactly or locally (in an inhomogeneous medium) descriptive of a plane wave whose propagation properties are determined by ξ. For example, in a plane stratified medium with z dependent refractive index $n(z)$ that varies slowly over a local-wavelength interval, the phase function is given typically by

$$\psi(r,r';\xi) = \xi(v-v') + \int_{z'}^{z} [n^2(\zeta) - \xi^2]^{1/2} d\zeta \qquad (1.75)$$

where v is a space coordinate (the wavenumber ξ along the v direction may alternatively be transformed into the propagation angle $w = \sin^{-1}\xi$). The direction of propagation of the local plane wave is given by $\nabla\psi$. Depending on the field representation employed and on whether the problem involves two or three space dimensions, it may be necessary to deal with a double integral over two wavenumber variables ξ and η, in which event A and ψ are also to be regarded as functions of ξ and η. With k assumed to be large[11], the integral may be evaluated by the saddle point method which requires deformation of the integration path P into "steepest descent paths" through the saddle points ξ_s of the integrand, due account being taken of any singularities (poles or branch points) that are encountered during the deformation. The saddle (or stationary) points are defined by the condition

$$d\psi(r,r';\xi)/d\xi = 0 \quad \text{at} \quad \xi_s, \qquad (1.76)$$

and yield upon solution $\xi_s = \xi_s(r,r')$. For example, in the plane stratified medium, one has

$$v - v' = \xi_s \int_{z'}^{z} [n^2(\zeta) - \xi_s^2]^{-1/2} d\zeta. \qquad (1.77)$$

The saddle point condition (1.76) is found to provide the equation for the ray trajectory that links the source point r' with the observation point r, possibly via a path that involves reflection at a boundary or diffraction at a scattering center (as noted previously, ray paths satisfy Fermat's principle of least time). The rays descriptive of a particular wave species (for example, the reflected or diffracted field), form a family or congruence, and ξ_s represents a parameter, the ray parameter, which distinguishes different rays in the family. The ray parameter is constant for all r and r' along a particular ray, i.e., $\xi_s(r,r') = C$, where C is an appropriate constant, is the equation of an entire ray. Correspondingly, $\xi_s(r_1,r')$ is the ray parameter for that ray which passes through r' and r_1 such that the rules for propagation, reflection, refraction,

---

[11] Alternatively, the observation distance may be taken as the large parameter and thus validate the asymptotic approximation also for lower frequencies.

or diffraction are satisfied. The asymptotic approximation to the integral in (1.74) is then given by[12]

$$I_\omega \sim I_{\omega s} + I_{\omega b} + I_{\omega p}, \tag{1.78}$$

where $I_{\omega s}$ is the saddle point contribution [1.17]

$$I_{\omega s} \sim V(r, r'; \xi_s) \{2\pi i [k \psi''(r, r'; \xi_s)]^{-1}\}^{1/2} \exp[ik\psi(r, r'; \xi_s)], \tag{1.79}$$

with $\psi'' \equiv (d^2\psi/d\xi^2)_{\xi_s}$ and ξ_s defined in (1.76). When ξ_s is constant, i.e., for observation points along a ray, the phase function $\psi(r, r'; \xi_s)$ is the same as the eikonal $S(r)$ in (1.60) (the dependence on the initial value r' on the ray trajectory is not indicated there), and the amplitude function V times the expression inside the braces is the same as $A(r)$ in (1.60). Note that the expression inside the braces accounts for the fact that $I_{\omega s}$ is synthesized not by the wave field $V \exp(ik\psi)$ at ξ_s alone but by a "bundle" or "packet" of waves with $\xi \approx \xi_s$ which interfere constructively at the observation point. Wave fields with ξ vastly different from ξ_s interfere destructively and hence do not contribute appreciably. By the saddle point method, the constructive interference is expressed as the change from the amplitude V to the "local" amplitude A[13]. When (1.76) can be solved explicitly for $\xi_s(r, r')$, the ray parameter may be eliminated from (1.79), which then provides an explicit dependence of the field on r and r'. Since ξ_s generally varies with r (except for r values on a single ray), different observation points are reached by local plane wave fields propagating in *different* directions.

The possible branch point and pole contributions I_b and I_p, respectively, encountered in the deformation of the original path into the steepest descent path, usually arise from singularities ξ_b and ξ_p that are fixed in the complex ξ-plane; i.e., they do not depend on r and r'. Thus, the corresponding eikonals $\psi(r, r'; \xi_{b,p})$ represent local plane wave fields propagating in the *same* direction, as described by ξ_b or ξ_p, toward any observation point. In a homogeneous medium, this property associates plane phase fronts with the corresponding wave fields. These

---

[12] It is implied in (1.78) that the critical points ξ_s, ξ_b, and ξ_p are isolated so that their contributions can be evaluated individually. In transition regions, where two or more critical points approach one another, it is necessary to employ a modified asymptotic procedure [1.17].

[13] The higher-order terms in inverse powers of $k = \omega/c$ in (1.62) may also be obtained from the saddle-point contribution.

features are illustrated by the integral representations for the Green's functions for a wedge and for a semi-infinite dielectric medium (Subsects. 1.5.1 and 1.5.3).

In the preceding discussion, it has been assumed that the critical points ξ_s, ξ_p or ξ_b are real, and that the associated ray trajectories in (1.76) and the phase functions $\psi(r,r';\xi_\alpha)\equiv S_\alpha(r,r')$, $\alpha=s$, p or b, are also real. This property characterizes the solutions $I_{\omega s}$ in (1.79), etc., as local *uniform (or homogeneous) plane wave* fields. When ξ_α and hence $\psi(r,r';\xi_\alpha)$ are complex, the resulting wave fields are local *nonuniform plane waves*, which travel along trajectories (still defined, for example, by (1.76) with $\xi_s(r,r')=$ constant) in a complex coordinate space ("complex rays"). Complex rays describe the propagation characteristics of local plane wave fields in dissipative media; they also characterize evanescent waves in non-dissipative media. Included in the evanescent wave category are surface wave fields exterior to open guiding structures, leaky wave fields, fields on the dark side of caustics generated by a focused incident field, and Gaussian beams. Some further observations on complex ray fields are made in Subsections 1.6.3 and 1.6.4 (for recent contributions, see [1.19]).

When the asymptotic results in (1.79), and correspondingly for I_b and I_p, are employed in the inversion formula (1.50), one may derive the transient field behavior near the various fronts as given in (1.63). If the propagation environment is dispersive, one may construct the transient field even at later observation times by retaining in (1.60) the dispersive eikonal $S(r,\omega)$ which, in a homogeneous medium, becomes $S=n(\omega)D$, where n is the refractive index and D is a propagation distance. It has already been shown that a high-frequency approximation as in (1.66) is adequate to provide the transition from the (non-dispersive) early-time wave front regime to the (dispersive) wave packet regime at moderately longer observation times. At still later times (i.e., at lower frequencies, with the observation distance chosen sufficiently large), it is appropriate to retain $n(\omega)$ intact without further approximation. The generic inversion integral then becomes

$$I \sim \frac{1}{2\pi}\int_{-\infty+i\varDelta}^{\infty+i\varDelta} A_\omega(r)\exp[i\varphi(r,\overline{r},\omega)]\,d\omega,\,\varphi(r,\overline{r},\omega)=(\omega/c_0)S(r,\overline{r},\omega)-\omega\tau$$

$$(1.80)$$

where $A_\omega(r)$ respresents an amplitude function that depends on r and ω, and it has been recognized that S generally involves also an initial coordinate $\overline{r}$ (for example, the source point $\overline{r}=r'$), As for the time-harmonic case in (1.74), the integral in (1.80) may be evaluated by the saddle point method. The saddle points $\omega_s(r, \bar{r}, \tau)$ in the complex ω plane are located by the solution of the equation

$$\frac{d}{d\omega}\left[\omega S(r, \bar{r}, \omega)\right] - c_0 \tau = 0 \quad \text{at} \quad \omega_s = \omega_s(r, \bar{r}, \tau), \qquad (1.81)$$

which defines trajectories in space-time, called space-time rays, traversed by wave packets (for a graphical solution of (1.81), see Fig. 1.5). The

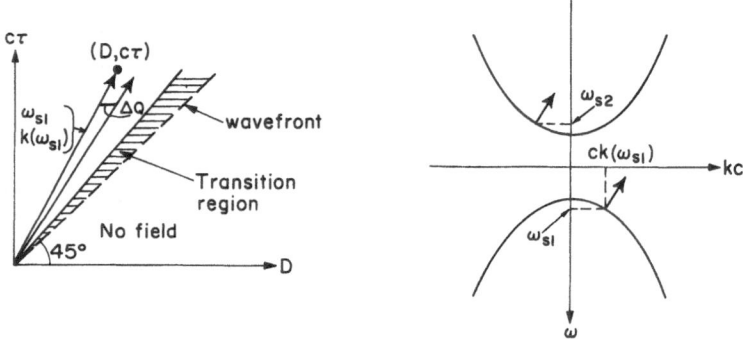

a) space – time ray plot   b) dispersion surface $k = k(\omega)$

Fig. 1.5a and b. Graphical solution of saddle point equation in a homogeneous medium. The vector from the origin to the observation point $(D, c\tau)$ in a) represents the space-time ray, whose slope $D/c\tau$ yields the relative group velocity v_g/c of the wavepacket. The wavepacket frequencies ω_{s1} and ω_{s2} are determined from the construction in b), which locates on the dispersion surface those points having normal vectors parallel to the space-time ray. The dispersion surface depicts the solutions of the dispersion equation $F(k, \omega) = 0$ or, equivalently, $k = k(\omega)$, with the wave-number k defined as $k(\omega) = \omega n(\omega)/c$, where n is the refractive index. The field amplitude in the wavepacket is proportional to the reciprocal of the square root of the ray tube cross section ΔQ

saddle point frequency ω_s represents the ray parameter (ω_s = constant on a space-time ray). In a homogeneous medium with $S = n(\omega)D$, (1.81) may be written as

$$\frac{D}{\tau} = v_g(\omega_s), \quad v_g(\omega) = c_0 \left\{\frac{d}{d\omega}\left[\omega n(\omega)\right]\right\}^{-1} \qquad (1.82)$$

where v_g is the group velocity. In an inhomogeneous medium, the group velocity is variable and is defined, for constant ω_s, by $dr/d\tau$

as derived from (1.81) [1.20]. The asymptotic approximation of the integral in (1.80) can be expressed as in (1.78) as

$$I \sim I_s + I_b + I_p \tag{1.83}$$

where $I_s = I_s^{(1)} + I_s^{(2)}$ is the saddle point contribution, with

$$I_s^{(1)} \sim A_{\omega_{s1}}(r)(-c_0/2\pi)^{1/2}\{i(d^2/d\omega^2)[\omega S(r,\bar{r},\omega)]_{\omega_{s1}}\}^{-1/2}\exp[i\varphi(r,\bar{r},\omega_{s1})],\tag{1.84}$$

$\omega_s \equiv \omega_{s1}$ being defined as a solution of (1.81)[14]. The field in (1.84) represents a wave packet which moves along a space-time ray with the constant frequency $\omega_s = \omega_s(r,\bar{r},\tau)$ and, in a homogeneous medium, with the constant velocity v_g. Since the saddle point frequency ω_s tracks an interference maximum, the group velocity represents the propagation speed of the energy in the wave packet. Alternatively, a stationary observer at r receives the field at different observation times in the form of wave packets with variable frequency $\omega_s(r,\bar{r},\tau)$. For a field with phase $\varphi[r,\bar{r},\omega_s(r,\bar{r},\tau)]$, this wave packet frequency is given by $\omega_s = -d\varphi/d\tau$, as may be ascertained for the present case by direct differentiation of φ in (1.84), recalling (1.81). Alternatively, the wave packet frequency is also given by $\omega_s = -\partial\varphi/\partial\tau$ when the dependence on ω_s and τ is given explicitly as in (1.80). The amplitude function multiplying the exponential in (1.84) accounts for the dispersive expansion or contraction of the wave packet and consequent decrease or enhancement of the field amplitude. The amplitude function is inversely proportional to the square root of the ray tube cross section in Fig. 1.5a. Thus, the diagrams in Figs. 1.5a and 5b give detailed information about the field behavior.

The branch point and pole contributions I_b and I_p in (1.83) arise when branch point or pole singularities at ω_b or ω_p are traversed during the deformation of the original integration path in (1.80) into the steepest descent paths through the saddle points. Branch points ω_b generally are located at the cutoff and resonance frequencies $\omega_b = \omega_c$ and $\omega_b = \omega_r$ in the medium, which characterize the transient field at long observation times and correspond to $n(\omega_c) = 0$ and $n(\omega_r) = \infty$, respectively. For a discussion of wave phenomena associated with various arrangements of critical points in the integrand of (1.80), see [1.18].

---

[14] Saddle points ω_s occur in pairs $\omega_{s1,2}$ such that the sum $I_s^{(1)} + I_s^{(2)}$ yields a real function. As in (1.79), the amplitude factor multiplying $A_{\omega_{s1}}$ in (1.84) accounts for the fact that the wave packet field is synthesized by wave constituents with a small frequency spread about the central frequency ω_{s1}.

Some of these phenomena are illustrated by an example in Subsection 1.6.2.

The space-time rays defined by $\omega_s(r, \bar{r}, \tau) = $ constant are real trajectories when ω_s is real; thereon, wave packets with dependence $\exp(i\omega_s S/c_0 - i\omega_s \tau)$ do not experience exponential attenuation. When the saddle points ω_s are complex, the corresponding exponentially attenuated wave packets move on complex space-time rays. This generalization is necessary to describe transient fields in lossy dispersive media; in lossless dispersive media, complex space-time rays arise when an input pulse has an exponentially varying amplitude profile (Gaussian, for example), or when a compressed input pulse is observed outside the strong-field region (in the shadow of the caustic formed by the converging space-time rays [1.21]).

Finally, it should be noted that time-dependent local plane wave fields of the form $B(r,t) \exp[i\varphi(r,t)]$ as in (1.84), where φ is a rapidly varying and B a slowly varying function, can be tracked directly through space-time without the prior utilization of the integral representation (1.80). This procedure, which utilizes the space-time ray equations, the dispersion equation for plane wave propagation in the medium, and transport equations for the field amplitude B, is a generalization of the ray method for time-harmonic fields. For a discussion of this method, see [1.20].

1.5 Pulse Diffraction in Non-Dispersive Media

The general considerations in Sections 1.3 and 1.4 are now illustrated by application to specific examples.

1.5.1 The Wedge

We consider the construction of the time-dependent scalar Green's function $G(r, r'; \tau)$, $\tau = t - t'$, for a perfectly conducting wedge as shown in Fig. 1.6. This requires solution of the time-dependent wave equation in (1.27), subject to the causality condition (1.28), to the edge condition (1.8a), and to an assumed Neumann boundary condition on the wedge faces: $\partial G/\partial \phi = 0$ at $\phi = 0, \Omega$. The problem is separable in a cylindrical (ϱ, ϕ, z) coordinate system and therefore admits of solution by eigenfunction expansion. The relevant one-dimensional completeness relations for a field representation as in Subsection 1.3.1 are [1.10]

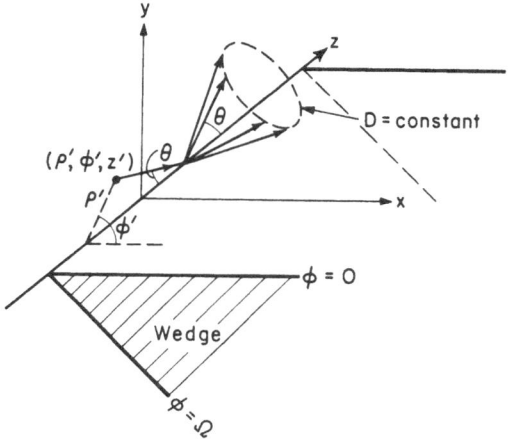

Fig. 1.6. Wedge geometry and cone of diffracted rays

$$\frac{\delta(\varrho - \varrho')}{\varrho'} = \int\limits_0^\infty \xi J_q(\xi \varrho) J_q(\xi \varrho') d\xi , \qquad 0 \le (\varrho, \varrho') < \infty \qquad (1.85)$$

$$\delta(\phi - \phi') = \frac{1}{\Omega} \sum\limits_{m=0}^\infty \delta_m \cos q\phi \cos q\phi' , \qquad 0 \le (\phi, \phi') \le \Omega , \qquad q = \frac{m\pi}{\Omega} \qquad (1.86)$$

$$\delta(z - z') = \frac{1}{2\pi} \int\limits_{-\infty}^\infty \exp[i\zeta(z - z')] d\zeta , \qquad -\infty < (z, z') < \infty \qquad (1.87)$$

$$\delta(\tau) = \frac{1}{2\pi} \int\limits_{-\infty + i\Delta}^{\infty + i\Delta} \exp(-i\omega\tau) d\omega, \qquad -\infty < \tau < \infty \qquad (1.88)$$

In (1.86), $\delta_m = 1$ for $m = 0$ and $\delta_m = 2$ for $m \ge 1$. Since the ϱ and z coordinates extend over an infinite interval, the corresponding eigenfunction spectra are continuous. Proceeding as in Subsection 1.3.1, one derives the representation

$$G(\boldsymbol{r}, \boldsymbol{r}'; \tau) = \frac{1}{(2\pi)^2 \Omega} \sum\limits_{m=0}^\infty \delta_m \int\limits_0^\infty d\xi \int\limits_{-\infty}^\infty d\zeta \int\limits_{-\infty + i\Delta}^{\infty + i\Delta} d\omega$$

$$\times \frac{\xi J_q(\xi \varrho) J_q(\xi \varrho') \cos q\phi \cos q\phi' \exp[i\zeta(z - z')] \exp(-i\omega\tau)}{\xi^2 + \zeta^2 - \omega^2/c^2}$$

$$(1.89)$$

which reveals the dispersion relation $\xi^2 + \zeta^2 = \omega^2/c^2$ to be satisfied for the eigenmodes in the wedge geometry.

The oscillatory eigenmode representation, as in Subsection 1.3.2, is obtained by following (1.42)–(1.44), or by residue evaluation at the poles $\omega^\pm$ of the ω-integral

$$G(\mathbf{r},\mathbf{r}';\tau) = \frac{c^2 U(\tau)}{2\pi\Omega} \sum_{m=0}^{\infty} \delta_m \int_0^\infty d\xi \int_{-\infty}^\infty d\zeta\, \xi J_q(\xi\varrho) J_q(\xi\varrho')$$

$$\cdot \cos q\phi \cos q\phi' \exp\left[i\zeta(z-z')\right] \frac{\sin\omega^+\tau}{\omega^+} \tag{1.90}$$

where $\omega^+ = c(\xi^2 + \zeta^2)^{1/2}$, and the Heaviside function $U(\tau)$ vanishes for $\tau < 0$. Because the wedge configuration is an open structure, the resonant frequencies $\omega^+ = -\omega^-$ are continuously distributed, in contrast to those in (1.40) for an enclosed region.

Referring to Subsection 1.3.3, there exist three guided eigenmode representations, depending on whether ϱ, ϕ or z is chosen as the preferred space coordinate. By following the procedure outlined in Subsection 1.3.3, or by residue evaluation at the poles $\zeta^\pm = \pm(\omega^2/c^2 - \xi^2)^{1/2}$ of the ζ-integral, one obtains for the z-guided representation (since $\mathrm{Im}\{\omega\} > 0$, one has $\mathrm{Im}\{\zeta^+\} > 0$, $\mathrm{Im}\{\zeta^-\} < 0$, whence the integration path in the complex ζ plane can be closed about ζ^+ when $(z-z') > 0$ and about ζ^- when $(z-z') < 0$):

$$G(\mathbf{r},\mathbf{r}';\tau) = \frac{i U(\tau)}{4\pi\Omega} \sum_{m=0}^{\infty} \delta_m \int_0^\infty d\xi \int_{-\infty+i\Delta}^{\infty+i\Delta} d\omega\, \xi J_q(\xi\varrho) J_q(\xi\varrho')$$

$$\cdot \cos q\phi \cos q\phi' \exp(-i\omega\tau) \frac{\exp\left[i\zeta^+ |z-z'|\right]}{\zeta^+}. \tag{1.91}$$

The ϱ-guided eigenmode representation may be derived from (1.89) by residue evaluation at the pole singularities $\xi^\pm = \pm(\omega^2/c^2 - \zeta^2)^{1/2}$ in the complex ξ plane, where $\mathrm{Im}\{\xi^\pm\} \gtrless 0$. To this end, it is convenient to change the semi-infinite ξ integral into an infinite integral by the transformation

$$\int_0^\infty d\xi\, \xi J_q(\xi\varrho) J_q(\xi\varrho') \ldots = \tfrac{1}{2} \int_{-\infty}^\infty d\xi\, \xi J_q(\xi\varrho_<) H_q^{(1)}(\xi\varrho_>) \ldots$$

where $\varrho_<$ and $\varrho_>$ denote the lesser and greater of the coordinates ϱ and ϱ' respectively. Then since $J_q(\xi\varrho_<) H_q^{(1)}(\xi\varrho_>)$ behaves as

$\xi^{-1} \exp[i\xi|\varrho-\varrho'|]$ as $\xi \to \infty$ in the upper half of the complex ξ plane, one may close the integration path about ξ^+ to obtain

$$G(\mathbf{r}, \mathbf{r}'; \tau) = \frac{iU(\tau)}{8\pi\Omega} \sum_{m=0}^{\infty} \delta_m \int_{-\infty}^{\infty} d\zeta \int_{-\infty+i\Delta}^{\infty+i\Delta} d\omega$$

$$\cdot \cos q\phi \cos q\phi' \exp[i\zeta(z-z')]\exp(-i\omega\tau)J_q(\xi^+\varrho_<)H_q^{(1)}(\xi^+\varrho_>). \tag{1.92}$$

Since the dispersion relation $\xi^2 + \zeta^2 = \omega^2/c^2$ in (1.89) does not involve the angular eigenvalue, the ϕ-guided eigenmode representation is most easily derived from (1.92) by converting the modal sum of angular functions into a contour integral in a complex $\bar{q}$ plane and then deforming the integration path about the singularities of the resulting integrand. Alternatively, this representation may be found directly by following for the ϕ coordinate the format outlined in (1.45)–(1.47), see also [1.22]. The result is

$$G(\mathbf{r}, \mathbf{r}'; \tau) = \frac{-U(\tau)}{8\pi^2} \int_{-\infty}^{\infty} d\zeta \int_{-\infty+i\Delta}^{\infty+i\Delta} d\omega \int_{-i\infty}^{i\infty} d\bar{q}\, q\, J_{\bar{q}}(\xi^+\varrho_<)H_{\bar{q}}^{(1)}(\xi^+\varrho_>)$$

$$\times \exp[i\zeta(z-z')]\exp(-i\omega\tau)\frac{\cos\bar{q}\phi_<\cos\bar{q}(\Omega-\phi_>)}{\bar{q}\sin\bar{q}\Omega} \tag{1.93}$$

where $\xi^+ = (\omega^2/c^2 - \zeta^2)^{1/2}$, and the principal value of the $\bar{q}$ integral is to be taken at $\bar{q}=0$. Moreover, to assure convergence of the field representation, the integration path at $\bar{q} = \pm i\infty$ is required to terminate at $|\arg\bar{q}| = \pi/2 - 0^+$, where 0^+ is a small positive quantity. The ϱ-guided eigenmode representation in (1.92) may be recovered from (1.93) by residue evaluation of the $\bar{q}$ integral at the pole singularities $\bar{q} = q = m\pi/\Omega$ in the right half of the complex $\bar{q}$ plane.

The formal field representations in (1.89)–(1.93) exhibit convergence properties which differ from one to the other and are therefore selectively useful for field evaluation in various parameter ranges. These representations may be further reduced by utilization of special integration and function-theoretic techniques. Returning to the oscillatory eigenmode representation in (1.90), one finds that the Fourier transform of $(\omega^+)^{-1}$ $\times \sin \omega^+ \tau$ can be evaluated in closed form and thereby permits elimination of the ζ integral [1.23,24]

$$c \int_{-\infty}^{\infty} d\zeta \exp[i\zeta(z-z')]\frac{\sin\omega^+\tau}{\omega^+} \tag{1.94}$$

$$= \pi J_0\{\xi[c^2\tau^2 - (z-z')^2]^{1/2}\} U(c\tau - |z-z'|).$$

Thus,

$$G(r, r'; \tau) = \frac{c\, U(c\tau - |z - z'|)}{2\Omega} \sum_{m=0}^{\infty} \delta_m \int_0^{\infty} d\xi\, \xi J_q(\xi \varrho) J_q(\xi \varrho')$$

$$\times \cos q\, \phi \cos q\, \phi'\, J_0\{\xi[c^2\tau^2 - (z - z')^2]^{1/2}\}. \tag{1.95}$$

Even the ξ integration can be performed explicitly [1.23]

$$\int_0^{\infty} d\xi\, \xi J_q(\xi \varrho) J_q(\xi \varrho') J_0(\xi c T)$$

$$= 0, \qquad\qquad\qquad 0 < cT < |\varrho - \varrho'|$$

$$= (\pi \varrho \varrho' \sin \varphi)^{-1} \cos q\, \varphi, \qquad |\varrho - \varrho'| < cT < \varrho + \varrho' \tag{1.96}$$

$$= \frac{(\sin q\pi) \exp(-q\beta)}{\pi \varrho \varrho' \sinh \beta}, \qquad \varrho + \varrho' < cT$$

where

$$\varphi = \cos^{-1}[(\varrho^2 + \varrho'^2 - c^2 T^2)(2\varrho\varrho')^{-1}],$$

$$\beta = \cosh^{-1}[(c^2 T^2 - \varrho^2 - \varrho'^2)(2\varrho\varrho')^{-1}]. \tag{1.97}$$

Thus, with $cT = [c^2\tau^2 - (z - z')^2]^{1/2}$ [1.24],

$$G(r, r'; \tau) = 0, \qquad\qquad\qquad\qquad 0 < cT < |\varrho - \varrho'|$$

$$= \frac{c}{2\pi \varrho \varrho' \Omega \sin \varphi} \sum_{m=0}^{\infty} \delta_m \cos q\, \phi \cos q\, \phi' \cos q\, \varphi,$$

$$\qquad\qquad\qquad\qquad |\varrho - \varrho'| < cT < \varrho + \varrho'$$

$$= \frac{c}{2\pi \varrho \varrho' \Omega \sinh \beta} \sum_{m=0}^{\infty} \delta_m \cos q\, \phi \cos q\, \phi' \sin q\pi \exp(-q\beta),$$

$$\qquad\qquad\qquad\qquad \varrho + \varrho' < cT. \tag{1.98}$$

While both series in (1.98) can be summed into a closed form expression, it should be noted that the second series is useful for field calculation when β is large, since the terms in the series then decrease very rapidly. Since β can be kept constant for appropriately scaled observation times and observation distances, one may have large β for any of the following parameter regimes, see (1.97): a) long observation times $(c^2\tau^2 \gg (\varrho + \varrho')^2 + (z - z')^2)$; b) observation point near the edge $(\varrho \to 0)$; c) source point near the edge $(\varrho' \to 0)$. For case a), $\beta \to \ln(c^2\tau^2/\varrho\varrho')$, whence the dominant $(m = 1)$ term in (1.98) yields

$$G \approx 2c(\Omega\pi)^{-1} \sin\frac{\pi^2}{\Omega} \cos\frac{\pi\phi}{\Omega} \cos\frac{\pi\phi'}{\Omega} (\varrho\varrho')^{\pi/\Omega} (c^2\tau^2)^{-1-\pi/\Omega}. \qquad (1.99)$$

Thus, the field for $cT \gg \varrho + \varrho'$, subsequently shown to be the diffracted field, decays monotonically. This behavior in a non-dispersive medium is to be contrasted with the long-time response in a dispersive environment, where oscillations occur at the natural resonance frequencies (see Sect. 1.6). Equation (1.99) also gives the correct dependence of the field on ϱ as $\varrho \to 0$, or on ϱ' as $\varrho' \to 0$, for arbitrary observation times after passage of the wave front (in these cases, one replaces $c^2\tau^2$ by $(c^2 T_1^2 - \varrho'^2)$ and $(c^2 T^2 - \varrho^2)$, respectively). The indicated behavior assures that vector fields derived from G, which involve both G and $\partial G/\partial\varrho$, satisfy the edge condition in (1.8a).

Returning to (1.98), one may eliminate the first series on use of the addition theorem for the product of two cosines and the formula

$$\sum_{m=0}^{\infty} \delta_m \cos q\phi_1 \cos q\phi_2 = \Omega\delta(\phi_1 - \phi_2). \qquad (1.100)$$

For $\Omega > \pi$, this leads to delta functions with the arguments $(|\phi \mp \phi'| - \varphi)$ and $(2\Omega - \phi - \phi' - \varphi)$. In the second series, one may express the trigonometric functions as exponentials and sum the resulting geometric series. The remarkable final closed form result is as follows (for $\Omega > \pi$ and $\phi' < \Omega - \pi$)[15]

$$G(\mathbf{r}, \mathbf{r}'; \tau) = \frac{\delta(\tau - |\mathbf{r} - \mathbf{r}'|/c)}{4\pi|\mathbf{r} - \mathbf{r}'|} U(\pi - |\phi - \phi'|) + \frac{\delta(\tau - |\mathbf{r} - \mathbf{r}_1'|/c)}{4\pi|\mathbf{r} - \mathbf{r}_1'|} U(\pi - \phi - \phi')$$

$$- \frac{c}{4\pi^2} \frac{\bar{Q}(\phi, \phi', \beta) + \bar{Q}(\phi, -\phi', \beta)}{\varrho\varrho'\sinh\beta} U(c\tau - D) \qquad (1.101)$$

where β is defined in (1.97) with $(cT)^2 = c^2\tau^2 - (z - z')^2$, $D = [(\varrho + \varrho')^2 + (z - z')^2]^{1/2}$, $\mathbf{r}_1'$ is the image of $\mathbf{r}'$ with respect to the $\phi = 0$ plane, and

$$\bar{Q}(\phi, \phi', \beta) = -\frac{\pi}{2\Omega} \left\{ \frac{\sin[(\pi/\Omega)(\phi - \phi' - \pi)]}{\cosh(\pi\beta/\Omega) - \cos[(\pi/\Omega)(\phi - \phi' - \pi)]} \right.$$

$$\left. - \frac{\sin[(\pi/\Omega)(\phi - \phi' + \pi)]}{\cosh(\pi\beta/\Omega) - \cos[(\pi/\Omega)(\phi - \phi' + \pi)]} \right\}. \qquad (1.102)$$

---

[15] Closed form solutions can also be derived for the case where $G=0$ on the wedge surface, and for the dyadic electromagnetic Green's functions [1.25].

The delta functions have here been expressed in an alternative form[16] that permits direct interpretation in terms of incident and reflected wave fronts since an impulsive point source in free space radiates a field

$$G_f(r, r'; \tau) = \frac{\delta(\tau - |r - r'|/c)}{4\pi |r - r'|},$$

(1.103)

i. e., a spherical wave front centered at the source point and propagating outward with velocity c. Due to the spherical spreading, the field strength decreases inversely with distance, but the field is confined to the wavefront shell $c\tau = |r - r'|$. The latter feature implies that when the temporal behavior of the source function is $h(t)$ for $t > 0$ instead of $\delta(\tau) = \delta(t - t')$, the corresponding field obtained on multiplication by $h(t')$ and integration from $t' = 0$ to $t' = \infty$, see (1.25), preserves the source pulse shape in the incident and reflected field response. This is not the case, however, for the diffracted field given by the third term in (1.101).

Referring to Fig. 1.7 one may interpret the result as follows. Before the incident pulse reaches the wedge face at $\phi = 0$ or the edge at $\varrho = 0$, the field is given by the first term in (1.101). When the incident wave

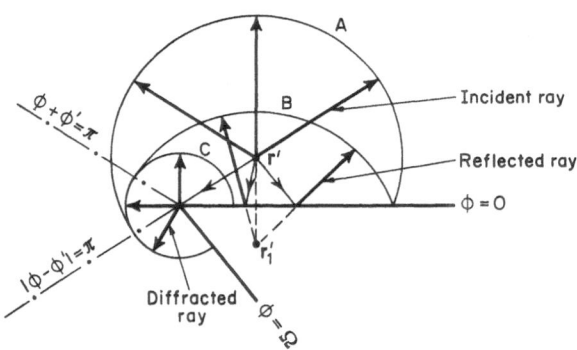

Fig. 1.7. Wavefronts and rays in $z = z'$ plane for wedge geometry. $A =$ incident wavefront: $c\tau = |r - r'|$; $B =$ reflected wavefront: $c\tau = |r - r_1'|$; $C =$ diffracted wavefront: $c\tau = \varrho + \varrho'$

front reaches the wedge face at $\phi = 0$, it produces a reflected wave front, the second term in (1.101), which appears to emanate from the source image and is therefore also spherical. When the incident wave front reaches the edge, it produces a diffracted wavefront located at

---

[16] Note that $\delta[f(\tau)] = [|df/d\tau|_{\tau_0}]^{-1} \delta(\tau - \tau_0)$, where τ_0 is a simple zero of $f(\tau)$.

$c\tau = D$ (third term in (1.101)), where D = constant defines the rim of a right circular cone whose axis coincides with the z axis, and whose apex is located such that the straight-line paths from the source point to the edge and from the edge to the observation point have identical angles θ with the edge (see Fig. 1.6). The straight-line paths, which lie on the cone, are the diffracted rays of the geometrical theory of diffraction. Since the observation points described by D = constant do not lie on a sphere, the diffracted wavefront is non-spherical. One observes from (1.101) that the diffracted field exists in all angular domains exterior to the wedge, and that the field persists after passage of the wave front, in contrast to the incident and reflected fields. The incident field exists only in the illuminated region $|\phi - \phi'| < \pi$ while the reflected field exists only in the region $(\phi + \phi') < \pi$, which defines the reflected wave domain according to geometrical optics. Note that the discontinuities across the wave front of the incident field are maintained during reflection but are weakened by diffraction. For $c\tau \approx D$, i.e., $\beta \approx 0$, the diffracted field in (1.101) becomes

$$G_{\mathrm{d}}(r, r'; \tau) \approx - \frac{c}{4\pi^2 (2D\varrho\varrho')^{1/2}(c\tau - D)^{1/2}} \left[\overline{Q}(\phi, \phi', 0) \right.$$
$$\left. + \overline{Q}(\phi, -\phi', 0) \right] U(c\tau - D), \qquad (1.104)$$

and thus exhibits near the diffracted wave front a behavior less violent than that of the delta function in the incident field[17]. The diffracted field has some of the characteristics associated with a cylindrical pulse, as may be seen from (1.127).

For very long observation times at a fixed observation point r, one finds from (1.102), with $\beta \to \ln(c^2\tau^2/\varrho\varrho')$,

$$\frac{\overline{Q}(\phi, \phi', \beta)}{\varrho\varrho' \sinh \beta} \to \frac{4\pi}{\Omega} (\varrho\varrho')^{\pi/\Omega} \sin \frac{\pi^2}{\Omega} \cos \left[\frac{\pi}{\Omega} (\phi - \phi') \right] (c\tau)^{-2-2\pi/\Omega}, c\tau \gg D, \qquad (1.105)$$

whence the resulting expression for G_{d} agrees with (1.99).

The result in (1.101) has been derived by starting from the oscillatory eigenmode representation in (1.90), and it has been possible to perform all integrations and summations explicitly. The various wave constituents (incident, reflected, and diffracted) and their wavefronts have been identified with the discontinuous behavior of the integrals in (1.94) and (1.96). The structure of the integrands, for example in (1.96), does not directly

---

[17] The behavior across the wavefront is also affected by the source function (see Subsect. 1.6.2).

make the resulting behavior evident. It is therefore of interest to explore alternative formulations that exhibit in a more systematic manner the separation of the total field into its constitutive parts. Since the field near the various wave fronts is synthesized by high-frequency spectral components, it may be anticipated that the desired representations should be well suited for the study of the high-frequency time-harmonic field[18]. This route, via time-harmonic analysis and subsequent Fourier inversion, also has the virtue of being more generally applicable since the reduction of field representations to a closed form is only rarely possible.

By comparing (1.50) with (1.91), (1.92), and (1.93), one may identify the various guided mode expansions of the time-harmonic Green's function G_ω. The ϕ-guided representation in (1.93) may be alternatively expressed in the following form [1.22].

$$G_\omega(r, r') = G_\omega^{(1)}(r, r') + G_\omega^{(2)}(r, r'), \tag{1.106}$$

where $G_\omega^{(1)}$ is the geometric-optical field, comprising incident and reflected spherical wave contributions, which are given for $\Omega > \pi$ and $\phi' < \Omega - \pi$ by

$$G_\omega^{(1)}(r, r') = \frac{\exp[ik|r - r'|]}{4\pi|r - r'|} U(\pi - |\phi - \phi'|)$$

$$+ \frac{\exp[ik|r - r'_1|]}{4\pi|r - r'_1|} U(\pi - \phi - \phi'), \tag{1.107}$$

with $k = \omega/c$. The remainder $G_\omega^{(2)}$ provides a rigorous correction to the geometrical-optics field and is given in the form of (1.55) with

$$Q(r, r'; w) = -\frac{i}{8\pi^2} [\bar{Q}(\phi, \phi', iw) + \bar{Q}(\phi, -\phi', iw)] [\gamma(r, r'; w)]^{-1} \tag{1.108}$$

$$\gamma(r, r'; w) = [D^2 + 2\varrho\varrho'(\cos w - 1)]^{1/2}, \tag{1.109}$$

where $\bar{Q}$ and D are defined in (1.102) and (1.101), respectively. Evidently, the time-dependent Green's function in (1.101) is recovered at once from (1.56), with the inversion of the geometric-optical portion performed directly via (1.50).

The high-frequency time-harmonic field can be evaluated by the saddle point method. The relevant saddle point w_s of (1.55), located

---

[18] Alternatively, a low-frequency approximation is useful for determining the long-time behavior as in (1.99).

at $d\gamma/dw = 0$, is seen from (1.109) to be $w_s = 0$. Thus, the integration path in (1.55) actually passes through the saddle point. The geometric-optical (incident and reflected) wave constituents are produced by pole singularities of $Q(r, r'; w)$ in the complex w-plane while the remainder, the diffracted field, is associated with the contribution from the saddle point integral. Thus, the various wave types are here clearly associated with the critical points (stationary points and singularities) in the integral representation. From (1.74) and (1.79), one finds for the high-frequency asymptotic approximation of the diffracted field,

$$G_\omega^{(2)} \sim -(32\pi^3)^{-1/2} \left[\bar{Q}(\phi, \phi', 0) + \bar{Q}(\phi, -\phi', 0) \right] \frac{\exp(ikD)}{(-ik\varrho'\varrho D)^{1/2}}, \quad k = \omega/c,$$

(1.110)

which can be interpreted in terms of the ray picture in Fig. 1.6. This simple result fails in the transition regions near the shadow and reflected wave boundaries $|\phi - \phi'| = \pi$ and $\phi + \phi' = \pi$, respectively, where the field amplitude diverges. Here, one requires an improved asymptotic evaluation that accounts for the proximity of a pole singularity and a saddle point in the integrand of (1.55).

When (1.110) is identified with the $m = 0$ term in (1.62), with $\alpha = -1/2$, one obtains via (1.63) and (1.65) the time-dependent field near the diffracted wavefront in (1.104), which was derived from the exact solution. The geometric-optical field in (1.107) can be identified with $A \exp(ikS)$ in (1.62) to yield via (1.63) and (1.65) the corresponding result in (1.101).

1.5.2 The Cone

The scalar and dyadic Green's functions for a perfectly conducting semi-infinite cone may be constructed by the method separation of variables and admit of alternative eigenmode expansions analogous to those in (1.90)–(1.93). However, it is not possible to obtain solutions in closed form as for the wedge problem. Therefore, one must here resort to approximate evaluation in various parameter ranges. For simplicity, we consider the scalar Green's function which satisfies (1.27) and (1.28) and vanishes on the cone surface $\theta = \theta_0$; here, (r, θ, ϕ) are spherical coordinates with origin at the cone tip. Also, it is assumed that the source is located on the cone axis (Fig. 1.8.), so that the field is independent of the azimuthal coordinate ϕ.

Recalling (1.98) for the wedge, which was found to be useful for calculation of the long-time response, it is suggestive to employ for

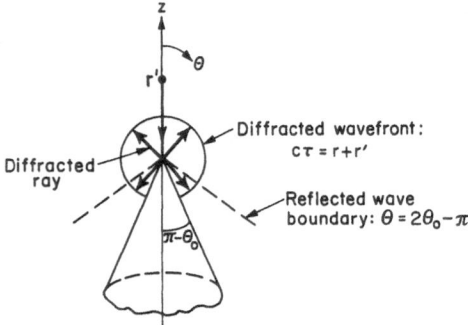

Fig. 1.8 Cone geometry, tip
diffracted rays and wavefront

the same parameter regime in the cone problem an expansion in θ-dependent eigenfunctions. The time-harmonic Green's function is given by [1.26]

$$G_\omega(r,r') = - \frac{\pi i}{8(rr')^{1/2}} \sum_q (2q+1) \frac{P_q(-\cos\theta_0)}{\sin q\pi(\partial/\partial q)P_q(\cos\theta_0)}$$

$$\cdot P_q(\cos\theta)J_{q+1/2}(kr_<)H^{(1)}_{q+1/2}(kr_>) \tag{1.111}$$

where $P_\nu(x)$ is the Legendre function of order ν and argument x, and the sum extends over all positive zeros q of $P_q(\cos\theta_0)=0$. This series converges rapidly when $(kr_<)$ is small since the Bessel function decays strongly for $(q+1/2)\gg kr_<$. The range $\omega r_</c\ll 1$ is accommodated either by low enough frequencies, for given source or observation point location, or by close enough proximity of the source or observation point to the cone tip, for given frequency. Translated into the time domain, one then expects the time-dependent Green's function representation obtained on Fourier or Laplace inversion of (1.111) to be rapidly convergent either for long observation times or for source or observation point locations near the cone tip.

The inversion may be accomplished on use of the formula [1.27]

$$\frac{1}{2\pi i} \int_{-i\infty+\Delta}^{i\infty+\Delta} I_\nu(sr_</c)K_\nu(sr_>/c)e^{s\tau}ds$$

$$= 0, \qquad\qquad\qquad\qquad c\tau < |r-r'|$$

$$= c(4rr')^{-1/2}P_{\nu-1/2}(\cos\varphi), \qquad |r-r'|<c\tau<r+r'$$

$$= c(\pi^2 rr')^{-1/2}\cos\nu\pi Q_{\nu-1/2}(\cosh\beta), \qquad c\tau>r+r' \tag{1.112}$$

where Q_ν is the Legendre function of the second kind, and

$$\varphi = \cos^{-1}\left[(r^2 + r'^2 - c^2\tau^2)(2rr')^{-1}\right],$$
$$\beta = \cosh^{-1}\left[(c^2\tau^2 - r^2 - r'^2)(2rr')^{-1}\right].$$

$$(1.113)$$

Also, I_ν and K_ν are modified Bessel and Hankel functions, respectively, defined as

$$I_\nu(x) = \exp(-i\nu\pi/2)J_\nu(ix), \quad K_\nu(x) = \frac{\pi i}{2}\exp(i\nu\pi/2)H_\nu^{(1)}(ix).$$

$$(1.114)$$

Thus, the time-dependent Green's function is, from (1.111) and (1.112),

$$G(r,r';\tau) = -\frac{c}{8rr'}\sum_q (2q+1)\frac{P_q(-\cos\theta_0)P_q(\cos\theta)}{\sin q\pi(\partial/\partial q)P_q(\cos\theta_0)}$$

$$\cdot \begin{cases} 0, & c\tau < |r-r'| \\ P_q(\cos\varphi), & |r-r'| < c\tau < r+r' \\ -\dfrac{2}{\pi}\sin q\pi\, Q_q(\cosh\beta), & c\tau > r+r'. \end{cases}$$

$$(1.115)$$

To interpret this result, we restrict observation angles to that region in space, which is not reached by fields reflected from the cone surface, i.e., $\theta < 2\theta_0 - \pi$. The expression in (1.115) for $|r-r'| < c\tau < r+r'$ may be recognized as [1.26]

$$G(r,r';\tau) = \frac{c}{4\pi rr'}\frac{\delta(\varphi - \theta)}{\sin\phi},$$

$$(1.116)$$

or, on changing variables from φ to τ (see footnote on p. 44),

$$G(r,r';\tau) = \frac{\delta(\tau - |r-r'|/c)}{4\pi|r-r'|},$$

$$(1.117)$$

where $|r-r'| = (r^2 + r'^2 - 2rr'\cos\theta)^{1/2}$. Thus, for $|r-r'| < c\tau < r+r'$, the observer sees only the incident pulse. At $\tau = (r+r')/c$, the observer is reached by the spherical diffracted wave front emanating from the cone tip and excited after the incident wave strikes the tip. The series in (1.115) for $c\tau > r+r'$, which yields the field after passage of the diffracted wave front, has so far not been summed into a closed form. However, for large τ, with $\beta \to \ln(c^2\tau^2/rr')$, one has the asymptotic approximation [1.28]

$$Q_q(\cosh \beta) \sim \frac{\pi}{2^{q+1}} \frac{\Gamma(q+1)}{\Gamma(q+3/2)} (\cosh \beta)^{-q-1} \qquad (1.118)$$

so that the series in (1.115) is rapidly convergent, as anticipated. From the leading term with $q = q_1$, one infers that the field decays according to τ^{-2q_1-2}. Since the dependence on r and r', as r or $r' \to 0$, is given correctly in (1.118), see remarks following (1.99), one notes the field behavior r^{q_1+1} near the cone tip. This assures that the electromagnetic fields derived from G obey the tip condition in (1.8b).

For efficient calculation of the high-frequency field, where kr and kr' are large, the series in (1.111) is unsuitable, and it is desirable to employ a contour integral representation, which can then be evaluated approximately by the saddle point method. Details of the calculation are given in [1.26]. It is found that outside the reflected field region $\theta < 2\theta_0 - \pi$,

$$G_\omega(r, r') \sim \frac{e^{ik|r-r'|}}{4\pi|r-r'|} + \left(\frac{e^{ikr'}}{4\pi r'}\right)\left[f(\theta)\frac{e^{ikr}}{(-ik)r}\right], \qquad k = \omega/c \qquad (1.119)$$

where the diffraction pattern function $f(\theta)$ is given in the form of an integral that can be approximated for small-angle cones as

$$f(\theta) \approx -\{2(1 + \cos\theta)\ln[(\pi - \theta_0)/2]\}^{-1}, \qquad \theta_0 \approx \pi. \qquad (1.120)$$

The second term in (1.119) has been factored so as to permit a direct ray-optical interpreation. The first factor represents the strength of the incident local plane wave field (here a spherical wave) at the cone tip. The second factor is the diffracted field when a plane wave of unit amplitude is incident along the z axis; this field has the form of a spherical wave centered at the tip, with angular amplitude distribution given by $f(\theta)$, and can be expressed in terms of tip diffracted rays as shown in Fig. 1.8. Referring to (1.62), one may deduce the transient field behavior at observation times near the incident and diffracted wave fronts

$$G(r, r'; \tau) \sim \frac{\delta[\tau - |r-r'|/c]}{4\pi|r-r'|} + \frac{cf(\theta)}{4\pi rr'} U\left(\tau - \frac{r+r'}{c}\right). \qquad (1.121)$$

One notes that the discontinuity in the field across the tip-diffracted wave front is bounded, in contrast to the edge diffracted field in (1.104).

Thus, the discontinuities across the incident wave front are weakened more by (three-dimensional) tip diffraction than by (two-dimensional) edge diffraction.

The above-described procedures may also be applied to the dyadic Green's functions, and to arbitrary source locations $(\theta' \neq 0)$ [1.26,29].

1.5.3 Dielectric Half Space

When a source is in the presence of a plane interface at $x=0$ separating two dielectric half spaces with (non-dispersive) propagation speeds c_1 and c_2, respectively, the incident wavefront excites a reflected wavefront in medium 1, where the source is located, and a refracted wavefront in medium 2. When $c_2 > c_1$, there exists in region 1 also a diffracted field, the lateral wave field, that is associated with the phenomenon of total reflection. In a lossless medium, and with the non-dispersive assumption, the transient fields can be constructed in simple form since the time-harmonic solution is representable as in (1.55). For example, for excitation by a z-directed electric line current located at (x', y') in medium 1, one obtains for the two-dimensional, z-independent scalar Green's function in medium 1 [1.30]

$$G(r,r';\tau) = G^{(1)}(r,r';\tau) + G^{(2)}(r,r';\tau) + G^{(3)}(r,r';\tau) \qquad (1.122)$$

where $G^{(1)}$ is the incident cylindrical wave field

$$G^{(1)}(r,r';\tau) = (2\pi)^{-1}(\tau^2 - |r-r'|^2/c_1^2)^{-1/2} U(c_1\tau - |r-r'|), \quad (1.123)$$

$G^{(2)}$ is the reflected field

$$G^{(2)}(r,r';\tau) = \frac{\mathrm{Re}\{\Gamma[\theta - i\cosh^{-1}(c_1\tau/|r-r_1'|)]\}}{2\pi(\tau^2 - |r-r_1'|^2/c_1^2)^{1/2}} U(c_1\tau - |r-r_1'|), \qquad (1.124)$$

and $G^{(3)}$ is the lateral wave diffracted field

$$G^{(3)}(r,r';\tau) = \frac{\mathrm{Im}\{\bar{\Gamma}[\cos^{-1}(c_1\tau/|r-r_1'|) + \theta]\}}{2\pi(|r-r_1'|^2/c_1^2 - \tau^2)^{1/2}} \qquad (1.125)$$

$$\times\ U(|r-r_1'| - c_1\tau)\,U[c_1\tau - |r-r_1'|\cos(\theta - \theta_c)]$$

Here, $r=(x,y)$, $r'=(x',y')$, $\theta_c = \sin^{-1}(c_1/c_2)$ is the critical angle for total reflection, θ is the angle between $|r-r'|$ and the normal to the interface, and $\Gamma(w)$ is the Fresnel reflection coefficient for a plane wave incident at the angle w,

$$\Gamma(w) = \{\cos w - [(c_1/c_2)^2 - \sin^2 w]^{1/2}\}\{\cos w + [(c_1/c_2)^2 - \sin^2 w]^{1/2}\}^{-1}. \tag{1.126}$$

$\bar{\Gamma}(w)$ is obtained from $\Gamma(w)$ on replacing $[(c_1/c_2)^2 - \sin^2 w]^{1/2}$ by $-i[\sin^2 w - (c_1/c_2^2)]^{1/2}$. The rays and wavefronts associated with the various wave contributions are schematized in Fig. 1.9. One observes that in contrast with the field excited by an impulsive point source, which

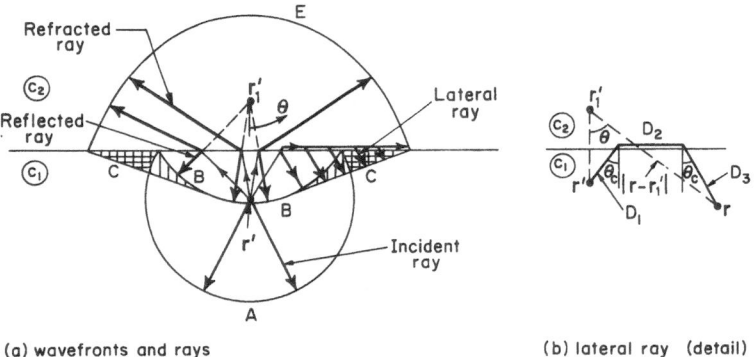

(a) wavefronts and rays   (b) lateral ray (detail)

Fig. 1.9a and b. Dielectric half space with $c_1 < c_2$. $A =$ incident wavefront: $c_1\tau = |r - r'|$; $B =$ reflected wavefront: $c_1\tau = |r - r'_1|$, where r'_1 is the location of the source image; $C =$ lateral wavefront: $c_1\tau = |r - r'_1|\cos(\theta_c - \theta)$, $\theta_c = \sin^{-1}(c_1/c_2)$ (lateral rays are shown only on one side); $E =$ refracted wavefront (non-circular). The lateral wave field exists only in the shaded region; observation points in the region with cross-hatched shading are reached first by the lateral wave

is confined to the wavefront, see (1.103), the field excited by an impulsive line source has a decaying tail behind the wavefront. The reflected wavefront appears to originate from an image source located at r'_1 in an infinite medium with propagation speed c_1. The laterally diffracted field for $\theta > \theta_c$ arises because the wavefront in region 2 travels faster than that in region 1 and therefore causes a spillover, by refraction, into region 1; the lateral wave field is absent when the source is located in the half space with the faster propagation speed.

The field discontinuities across the various wavefronts are readily ascertained. From (1.123), one has for the incident field,

$$G^{(1)} \approx (2\pi)^{-1}(2|r-r'|/c_1)^{-1/2}(\tau - |r-r'|/c_1)^{-1/2}, \quad c_1\tau \approx |r-r'|,$$
$$(1.127)$$

a less violent behavior than that in (1.103) for the spherical pulse. The reflected field near the wavefront,

$$G^{(2)} \approx (2\pi)^{-1}[\text{Re}\{\Gamma(\theta)\}](2|r-r_1'|/c_1)^{-1/2}(\tau - |r-r_1'|/c_1)^{-1/2},$$
$$c_1\tau \approx |r-r_1'|,$$
$$(1.128)$$

behaves like the incident field in (1.127) but with an observation-angle-dependent amplitude. Near the lateral wavefront $c_1\tau \approx |r-r_1'|\cos(\theta-\theta_c)$, one has $w \approx \theta_c$ whence $\text{Im}\{\bar{\Gamma}\} \to 0$. Therefore, the inverse cosine in (1.125) must be expanded to the next order of approximation. The result,

$$G^{(3)} \approx \frac{(\sin 2\theta_c)^{1/2}}{\pi\cos\theta_c[(|r-r_1'|/c)\sin(\theta-\theta_c)]^{3/2}}[\tau - (|r-r_1'|/c)\cos(\theta-\theta_c)]^{1/2}, \theta > \theta_c,$$
$$(1.129)$$

reveals substantial smoothing of the incident field discontinuity across the lateral wavefront. When (1.127)–(1.129) are employed in (1.62) and (1.63), one obtains for the high-frequency time-harmonic field,

$$G_\omega^{(1)} \sim [8\pi|r-r'|/c_1]^{-1/2}(-i\omega)^{-1/2}\exp[i(\omega/c_1)|r-r'|] \qquad (1.130)$$

$$G_\omega^{(2)} \sim [8\pi|r-r_1'|/c_1]^{-1/2}\Gamma(\theta)(-i\omega)^{-1/2}\exp[i(\omega/c_1)|r-r_1'|] \quad (1.131)$$

$$G_\omega^{(3)} \sim \frac{(\sin 2\theta_c)^{1/2}c^{3/2}}{2\pi^{1/2}\cos\theta_c[|r-r_1'|\sin(\theta-\theta_c)]^{3/2}}$$
$$\cdot\frac{\exp[(i\omega/c_1)|r-r_1'|\cos(\theta-\theta_c)]}{(-i\omega)^{3/2}}, \qquad \theta > \theta_c.$$
$$(1.132)$$

The phase in (1.132) describes a field travelling from the source to the interface along the critically incident ray, along the lateral ray in the second medium, and back into the first medium along a ray produced by lateral shedding, as may be seen from the identity

$$|r-r_1'|c_1^{-1}\cos(\theta-\theta_c) = c_1^{-1}D_1 + c_2^{-1}D_2 + c_1^{-1}D_3, \qquad (1.133)$$

with $D_{1,2,3}$ defined in Fig. 1.9b. Also, one may write

$$|r-r_1'|\sin(\theta-\theta_c) = (c_1/c_2)[1-(c_1/c_2)^2]^{1/2}D_2. \qquad (1.134)$$

For observation angles in the transition region $\theta \approx \theta_c$ (i.e., $D_2 \approx 0$), the high-frequency asymptotic solution must be modified.

When the dielectrics are dissipative and dispersive, closed form solutions for the transient field are generally not attainable (however, an exception occurs for a lossless plasma [1.31]). Solutions must then be obtained by asymptotic methods as in Subsection 1.4.5 or by numerical procedures [1.32,33]. For the case of a realistically modeled ground environment, such procedures are discussed in Chapter 5.

1.6 Pulse Propagation in Dispersive Media

1.6.1 Lossless Isotropic Plasma—Impulse Excitation

To observe how dispersion affects the propagation characteristics of pulsed input signals, we consider first the simple example of a cold lossless, isotropic, homogeneous, unbounded plasma medium, for which the scalar Green's function is defined by (1.30), with (1.28). For further simplification, it is assumed that the excitation is by a uniform line current located along the z axis so that the field is independent of z. Because of rotational symmetry, the field is a function only of $\varrho = (x^2 + y^2)^{1/2}$, with $\varrho' = 0$. The representation of G by the various eigenfunction expansions discussed in Section 1.3 can be deduced from Subsection 1.5.1 if the angular eigenfunctions in (1.86) for the wedge region are replaced by functions with (2π) periodicity; i.e., $q \equiv m$, $\Omega = 2\pi$, and $\phi' = 0$. The uniform line current along the z axis can be synthesized by superposition of point sources from $z' = -\infty$ to $z' = +\infty$; the resulting integration introduces into the integrand of (1.89) the delta function $2\pi\,\delta(\zeta)$, thereby removing the ζ-integration and implying an evaluation of the integrand at $\zeta = 0$. Finally, since $\varrho' = 0$, the only non-vanishing contribution to (1.85) and (1.86) comes from the $m = 0$ term, thereby reducing the sum in (1.89) to the single term $(1/2\pi)$.

The above-noted modifications yield various representations for the line source field $G(\varrho, \tau)$ in a non-dispersive medium as defined by (1.27) and (1.28). To account for dispersion as in (1.30), the only modification is the inclusion of ω_p^2/c^2 in the denominator of the reduced form of (1.89)

$$G(\varrho, \tau) = \frac{1}{(2\pi)^2} \int\limits_0^\infty d\xi \int\limits_{-\infty + i\varDelta}^{\infty + i\varDelta} d\omega \, \frac{\xi J_0(\xi \varrho) \exp(-i\omega\tau)}{\xi^2 - (\omega^2 - \omega_p^2)c^{-2}}, \qquad \tau = t - t'.$$
$$(1.135)$$

Then instead of (1.90),

$$G(\varrho,\tau) = \frac{c^2 U(\tau)}{2\pi} \int\limits_0^\infty d\xi \, \xi J_0(\xi \varrho) \frac{\sin(\omega^+ \tau)}{\omega^+}, \tag{1.136}$$

where $\omega^+ = (c^2\xi^2 + \omega_p^2)^{1/2}$, and instead of (1.92),

$$G(\varrho,\tau) = \frac{i U(\tau)}{8\pi} \int\limits_{-\infty + i\Delta}^{\infty + i\Delta} d\omega \, H_0^{(1)}(\xi^+ \varrho) \exp(-i\omega\tau) \tag{1.137}$$

where $c\xi^+ = (\omega^2 - \omega_p^2)^{1/2}$. The spatial and temporal periodicities of the oscillatory and radially guided eigenmodes in (1.136) and 1.137) satisfy the dispersion equation $\xi^2 c^2 = \omega^2 - \omega_p^2$. The integrals can be evaluated in closed form [1.28]

$$G(\varrho,\tau) = \frac{\cos[\omega_p(\tau^2 - \varrho^2/c^2)^{1/2}]}{2\pi(\tau^2 - \varrho^2/c^2)^{1/2}} U(c\tau - \varrho). \tag{1.138}$$

Several observations follow from this result. First, when $\omega_p = 0$, (1.138) reduces properly to the solution (1.123) in a non-dispersive environment. Near the wavefront at $\tau = \varrho/c$, the effect of dispersion is negligible and the field discontinuity is described by (1.127). However, at long observation times,

$$G(\varrho,\tau) \sim \frac{\cos \omega_p \tau}{2\pi\tau}, \tag{1.139}$$

so that the field oscillates with decreasing amplitude at the plasma frequency ω_p. At arbitrary observation times, one may define the local oscillation frequency ω_s as the time derivative of the phase

$$\omega_s = (d/d\tau)[\omega_p(\tau^2 - \varrho^2/c^2)^{1/2}] = \omega_p \tau(\tau^2 - \varrho^2/c^2)^{-1/2}, \tag{1.140}$$

and observe that ω_s decreases monotonically from infinity at the wavefront to ω_p far behind the wavefront. The local wavenumber k descriptive of the spatial period of the oscillatory field may be defined as the spatial derivative of the phase whence

$$k_s = \varrho \omega_p c^{-2}(\tau^2 - \varrho^2/c^2)^{-1/2}. \tag{1.141}$$

The local wavelength $2\pi/k_s$ therefore increases from zero near the wavefront to infinity far behind the wavefront. However, both ω_s and k_s

remain constant when the observer moves with constant speed ϱ/τ. Since infinite wavelength is indicative of the absence of wave motion, no energy transport is associated with the long-time oscillations at the plasma frequency. The frequency and wavenumber satisfy the dispersion equation $k_s^2 c^2 = \omega_s^2 - \omega_p^2$, and the phase may be written alternatively as $(k\varrho - \omega t)$.

When the evaluation of the integral in (1.137) is performed by the asymptotic procedure of Subsection 1.4.5, one first approximates the time-harmonic field by its large argument form $H_0^{(1)}(\alpha) = (2/\pi\alpha)^{1/2}$ $\exp(i\alpha - i\pi/4)$, whence the phase function ωS in (1.80) becomes $(c\xi^+\varrho)$ and the amplitude function A_ω becomes $(-8i\pi\varrho\xi^+)^{-1/2}$. The saddle points $\omega_{s1,2}$ in (1.81) are then given by $\pm\omega_s$ in (1.140) (see also the graphical construction in Fig. 1.5), and the steepest descent paths through the saddle points are shown in Fig. 1.10. Since there are no singularities between P and P', the asymptotic approximation to the integral in

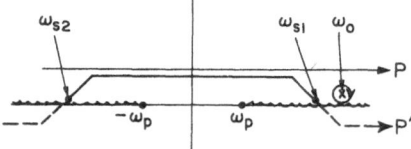

Fig. 1.10. Integration paths and singularities in complex ω plane for pulse propagation in lossless plasma medium. $P =$ original path, $P' =$ deformed path through saddle points $\omega_{s1,2}$. Branch points at $\pm\omega_p$ are due to ξ^+ in (1.137); $\xi^+ > 0$ and $\xi^+ < 0$ on the top side of the branch cut to the right and left, respectively. The pole at ω_0 arises for the excitation in (1.142) and is absent for impulsive excitation. The dashed portion of P' proceeds on the lower Riemann sheet

(1.137) is obtained directly from (1.84). The result is found to be given by (1.138) but now the restriction $|\xi^+(\omega_s)\varrho| \equiv |k_s\varrho| \gg 1$ must be imposed in order to validate the above-noted simplification of the integrand in (1.137). Also, (1.84) requires $|\omega_s| < \infty$ since $(d^2/d\omega^2)(\omega S) \to 0$ as $\omega \to \infty$. Thus, the asymptotic result does not generally accommodate the very short and very long time ranges although it happens to do so in the present special case. For observation times near the wavefront, one uses (1.66) with $\omega_1^2 = \omega_p^2/2$; since $A_\omega = (-8i\pi\varrho\xi^+)^{-1/2} \to (8\pi\varrho/c)^{-1/2}$ $(-i\omega)^{-1/2}$, one identifies $u_m(r)$ in (1.71) with the first factor and $\xi = 1/2$. Eq. (1.68) then provides the transition from the wavepacket regime (1.71) (which agrees with (1.138) as $\tau \to \varrho/c$) to the wavefront regime (1.69) (which agrees with (1.127)). For long observations times, when ω_s approaches the branch points $\pm\omega_p$, see [1.18].

1.6.2 Lossless Isotropic Plasma—Harmonic Step Signal

To assess the propagation characteristics of input pulses other than the idealized impulse, we consider the time function

$$h(t) = \exp(-i\omega_0 t) U(t) \tag{1.142}$$

which describes a suddenly initiated harmonic signal with constant frequency ω_0. By (1.54), the Fourier transform is $h(\omega) = i(\omega - \omega_0)^{-1}$, and the corresponding response $G(\varrho, t)$ is obtained from (1.137) on replacing τ by t and including $h(\omega)$ in the integrand. This modification introduces a pole at $\omega = \omega_0$ but does not otherwise alter the saddle point and singularity configuration in the integrand. Consequently, the asymptotic evaluation of the integral proceeds along the path P' in Fig. 1.10 as before, but now the deformation of P into P' differs depending on whether $\omega_{s1} > \omega_0$ or $\omega_{s1} < \omega_0$. In the latter case, the pole is traversed during the path deformation and requires extraction of a residue contribution (I_p in (1.83)). The result for ($I_s + I_p$) in (1.83) is

$$G(\varrho, t) \sim G_1(\varrho, t) + G_2(\varrho, t), \tag{1.143}$$

where $G_1(\varrho, t)$ is the residue contribution

$$G_1(\varrho, t) = \frac{i}{4} H_0^{(1)} \big[(\varrho/c)(\omega_0^2 - \omega_p^2)^{1/2} \big] \exp(-i\omega_0 t) U(t - t_0) \tag{1.144}$$

with

$$t_0 = (\varrho/c)(\omega_0^2 - \omega_p^2)^{-1/2} \omega_0 = \varrho/v_{g0} \tag{1.145}$$

denoting the observation time in (1.140) when the saddle point ω_s coincides with the pole ω_0. The saddle point contribution for $t > \varrho/c$ is

$$G_2(\varrho, t) = \frac{i}{4\pi \bar{t}} \left[\frac{\exp(-i\omega_p \bar{t})}{\omega_s - \omega_0} - \frac{\exp(i\omega_p \bar{t})}{\omega_s + \omega_0} \right], \quad \bar{t} = \left(t^2 - \frac{\varrho^2}{c^2} \right)^{1/2}, \tag{1.146}$$

where $\omega_s = \omega_p t/\bar{t}$ as in (1.140).

$G_1(\varrho, t)$ is recognized as the steady-state field excited by a harmonic line source in a medium with refractive index $n(\omega) = (1 - \omega_p^2/\omega^2)^{1/2}$. The residue contribution has been given exactly in (1.144) without use of the large argument approximation of the Hankel function. The arrival time t_0 of the harmonic signal at the observation point ϱ corresponds to the wavepacket group velocity $v_{g0} \equiv v_g(\omega_0)$ defined in (1.82). Because ω_0 is constant, wave packets in the harmonic signal continue to arrive at all subsequent times with the same velocity (see. Fig. 1.11). When $\omega_0 < \omega_p$, the refractive index is positive imaginary and the Hankel func-

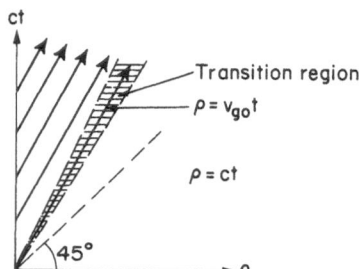

Fig. 1.11. Space-time rays for time-harmonic field in (1.144). The frequency on all rays is ω_0 and the group velocity is v_{g0} in (1.145)

tion in (1.144) is exponentially small; the medium is opaque to waves below the cutoff frequency ω_p. To within the asymptotic approximation of the total field, G_1 may then be neglected (when $\omega_0 < \omega_p$, the pole is not intercepted during the path deformation in Fig. 1.10). $G_2(\varrho, t)$ in (1.146) is the transient response, which exists for all $t > \varrho/c$. The result is not valid near the wavefront $t = \varrho/c$, near the signal arrival time t_0 where $\omega_s \rightarrow \omega_0$, and at very long observation times (for the required transition functions in the latter two regimes, see [1.18]). The space-time ray diagram is composed of radial trajectories as in Fig. 1.5. The space-time ray diagram for the composite field is obtained by combining Figs. 1.5 and 1.11.

Various special cases can be derived from (1.143). The response to the real time signal

$$h(t) = U(t)\cos\omega_0 t \tag{1.147}$$

is obtained from ReG:

$$\mathrm{Re}\{G\} \sim \frac{\sin\omega_p \bar{t}}{2\pi} \frac{\omega_p t}{(\omega_p t)^2 - (\omega_0 \bar{t})^2} + \mathrm{Re}\{G_1(\varrho, t)\} \tag{1.148}$$

with its evident breakup into a steady-state and transient portion. When $\omega_0 = 0$, the excitation degenerates into the dc step

$$h(t) = U(t) \tag{1.149}$$

whence upon omission of the exponentially small Hankel function in (1.144),

$$\text{Re}\{G\} \sim \frac{\sin \omega_p \bar{t}}{2\pi \omega_p t}, \quad \omega_0 = 0. \tag{1.150}$$

For long observation times, the transient solutions in (1.148) or (1.150) decay like t^{-1} and oscillate at the plasma frequency, similar to the impulse-excited field in (1.139). Near the wavefront $t = \varrho/c$, one may show from the high-frequency time-harmonic solution and application of (1.63) that for the source functions in (1.147) or (1.149)

$$\text{Re}\{G\} \sim \pi^{-1}(c/2\varrho)^{1/2}(t - \varrho/c)^{1/2}, \quad t \approx \varrho/c. \tag{1.151}$$

This behavior is consistent with that derived from (1.127) by integration over t' to generate a unit step input function from the impulse $\delta(\tau) = \delta(t - t')$. The absence of ω_0 in (1.151) is indicative of the fact that in the excitation function (1.142), the initial discontinuity which governs the behavior across the wave-front is independent of ω_0, as it is in (1.149). One observes also that a more regular source behavior at $t = 0$ produces a smoother field variation across the wavefront.

The response to the source function

$$h(t) = U(t) \sin \omega_0 t \tag{1.152}$$

is obtained from $\text{Re}\{iG\}$ in (1.143)

$$\text{Re}\{iG\} \sim \frac{\cos \omega_p \bar{t}}{2\pi} \frac{\omega_0 \bar{t}}{(\omega_0 \bar{t})^2 - (\omega_p t)^2} + \text{Re}\{iG_1(\varrho, t\}. \tag{1.153}$$

The general features of this solution are similar to those in (1.148) but the field near the wavefront, as ascertained from the high-frequency time-harmonic behavior of G_ω and application of (1.63), is

$$\text{Re}\{iG\} \sim (\omega_0/3\pi)(2c/\varrho)^{1/2}(t - \varrho/c)^{3/2}, \quad t \approx \varrho/c, \tag{1.154}$$

more regular than that in (1.151). The field now depends on ω_0, which provides a measure of the discontinuity in the derivative of the continuous source function (1.152) at $t=0$.

1.6.3 Lossless Isotropic Plasma—Half Space

When a pulsed source is located exterior to a dispersive semi-infinite medium, the phenomena described in Subsection 1.5.3, where the propagation speeds $c_{1,2}$ are assumed to be frequency independent, are modified. For a strongly dispersive medium with $c_2 = c_2(\omega)$, the refractive index $n(\omega) = c_1/c_2(\omega)$, where c_1 is a constant, may be characterized at high frequencies by $(1 - \omega_i^2/\omega^2)$, as in (1.66). This does not alter the incident wave field in (1.130). However, the reflection coefficient $\Gamma(\theta)$ as given by (1.126) tends to zero as $\omega \to \infty$ since $c_1/c_2 = n \to 1$. It is therefore necessary to expand $\Gamma(\theta)$ in inverse powers of ω^2, and the leading term varies as $\Gamma = b\omega^{-2}$, where b is a frequency independent coefficient. Thus, $G_\omega^{(2)}$ in (1.131) depends on ω as $\omega^{-5/2}$, whence the behavior near the reflected wavefront in (1.128) is changed from $(\tau - |r - r_1'|/c_0)^{-1/2}$ to $(\tau - |r - r_1'|/c_0)^{3/2}$. The field discontinuity across the reflected wavefront, which is still centered at the image point, is therefore weakened substantially since the strongly dispersive half space is transparent to high-frequency waves. Moreover, since $\theta_c \to \pi/2$, the lateral wavefront is not excited.

At later observation times, but at large observation distances, one may employ in (1.80) the asymptotic results in (1.130)–(1.132) provided that the refractive index $n(\omega) = c_1/c_2(\omega)$ is not approximated by its high-frequency form. The incident and reflected field integrals cannot now be evaluated by the saddle point method because the exponentials are of the form $\exp(iq\omega)$, where q is independent of ω. However, the lateral wave integral gives rise to a dispersive wave process. The stationary points ω_s are defined from (1.81) by, see (1.133),

$$\tau = D_1/c_1 + D_2/v_{g2}(\omega_s) + D_3/c_1, \qquad v_{g2}(\omega) = c_1 \left\{ \frac{d}{d\omega} [\omega n(\omega)] \right\}^{-1}, \tag{1.155}$$

and thus select those wave frequencies compatible with the travel time τ along the lateral ray trajectory shown in Fig. 1.9b (i.e., $\sin \theta_c = n(\omega_s)$). Since the group propagation speeds along the different path segments appear in (1.155) (in region 1, $v_{g1} = c_1$), the saddle point does indeed track the wave group that delivers field energy to the observation point r. The transient field may then be evaluated from (1.84). For the special case of a cold isotropic plasma with $n^2 = (1 - \omega_p^2/\omega^2)$, which also admits of a closed form solution, see [1.31].

1.6.4 Gaussian Pulses in Lossless Media: Complex Space-Time Rays

When a time-harmonic input pulse has a Gaussian envelope,

$$h(t) = \exp(-i\omega_0 t - t^2/4\alpha^2), \qquad \alpha > 0, \tag{1.156}$$

the spectrum function $h(\omega)$ is also Gaussian,

$$h(\omega) = \alpha \pi^{-1/2} \exp[-\alpha^2(\omega - \omega_0)^2]. \tag{1.157}$$

Insertion of $h(\omega)$ into the integrand of (1.137) and replacement of τ by t then yields the transient response $G(\varrho, t)$. For an asymptotic evaluation of the integral at large ϱ, one replaces the Hankel function by its large argument approximation, whence the phase function φ in (1.80) is given by[19]

$$\varphi(\varrho, \omega) = k(\omega)\varrho - \omega t + i\alpha^2(\omega - \omega_0)^2. \tag{1.158}$$

Thus, φ is complex for all real ω.

For a very wide pulse that contains many cycles of the harmonic carrier, one has $\alpha \gg 1$ and hence a frequency spectrum narrowly confined about $\omega = \omega_0$. One may then expand A_ω and $k(\omega)$ in (1.80) in a power series about $\omega = \omega_0$ and effect the Fourier inversion in terms of known integrals [1.34]. For very narrow pulses where $\alpha \ll 1$, one may treat $\exp[-\alpha^2(\omega - \omega_0)^2]$ as part of the amplitude function A_ω, look for saddle points of the remaining real phase, and perform the asymptotic evaluation as previously. However, when these assumptions are inapplicable, the phase φ in (1.158) must be retained intact, and the saddle points ω_s defined by $d\varphi/d\omega = 0$, viz.,

$$\omega_s = \omega_0 - (i/2\alpha^2)[t - \varrho(dk/d\omega)_{\omega_s}], \qquad \omega_s = \omega_s(\varrho, t), \tag{1.159}$$

are complex. Hence, the phase $\varphi(\varrho, \omega_s)$ in the asymptotic solution (1.84), is also complex, and the transient field has an exponential amplitude dependence.

---

[19] For the present discussion, $\xi^+(\omega)$ may be an arbitrary dispersive propagation function in a lossless medium and is therefore replaced by $k(\omega)$.

The complex value of the group velocity $v_g = (dk/d\omega)^{-1}$ makes it difficult to associate this quantity with the energy transport properties of the signal. However, one may investigate the propagation characteristics of the pulse maximum, defined (for slowly varying amplitude functions in (1.84)) by the temporal or spatial maxima of $\mathrm{Im}\{\varphi\}$. The condition $(d/dt)\mathrm{Im}\{\varphi\} = 0$ locates at a given observation point ϱ the temporal maximum of the field while the condition $(d/d\varrho)\mathrm{Im}\{\varphi\} = 0$ locates at a given time t the spatial maximum. Since from (1.158) and (1.159),

$$d\varphi(\varrho,\omega_s)/dt = -\omega_s + [(dk/d\omega_s)\varrho - t + i2\alpha^2(\omega_s - \omega_0)]d\omega_s/dt = -\omega_s,$$
$$(1.160)$$

the frequency ω_s is real at the temporal maximum. Noting that $dk/d\omega$ for a lossless medium is real when ω is real, one has from (1.159),

$$\omega_s = \omega_0 \quad \text{at the temporal maximum}, \tag{1.161}$$

and

$$\varrho/t = v_g(\omega_0) \tag{1.162}$$

as the equation of motion of the temporal maximum. Thus, the temporal maximum of the exponential portion of the pulse field retains the carrier frequency ω_0 and moves past a stationary observer at any point ϱ with the group speed of the carrier frequency, i.e., along a real space-time ray.

The condition for the spatial maximum, $(d/d\varrho)\mathrm{Im}\{\varphi\} = 0$, may be shown to imply that $k(\omega_s)$ is real. It then follows that (1.161) and (1.162) are again satisfied so that the equations of motion of the spatial and temporal peaks are identical, with constant frequency ω_0. Therefore, one may still associate the group velocity $v_g(\omega_0)$ with a Gaussian wave packet in a lossless medium. This is not the case when losses are present (see Subsect. 1.6.5).

At other points on the Gaussian envelope, the saddle point frequencies are complex and change with time. Such points do not follow straight trajectories in space-time although the medium is homogeneous. The tracking of wave packets with complex frequency generally requires use of "complex space-time rays", i.e., trajectories in a complex coordinate space [1.35]. The physical fields at real (ϱ, t) are then obtained from the intersections of the complex rays with real space. To understand why complex rays are required, consider the source function $h(t')$ in

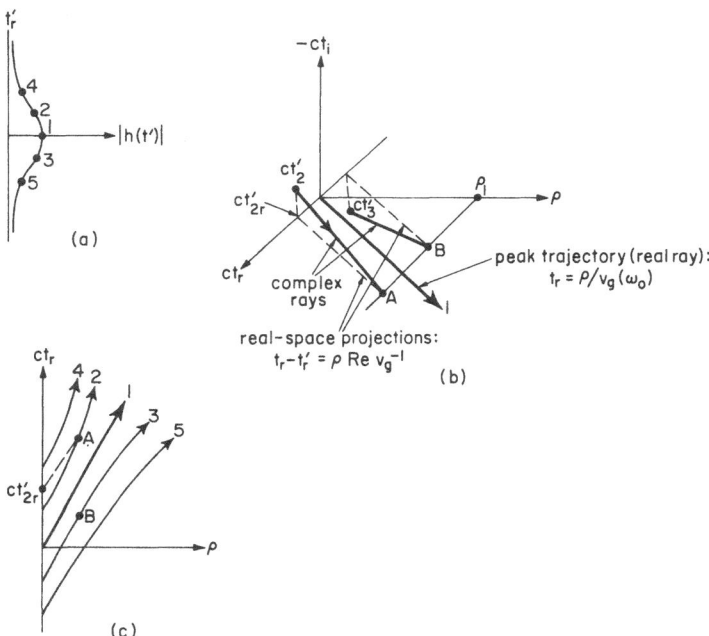

Fig. 1.12a–c. Complex space-time rays for propagation of a Gaussian pulse with harmonic carrier in a lossless homogeneous medium. $t_r = \mathrm{Re}\{t\}$, $t_i = \mathrm{Im}\{t\}$, $\varrho = \mathrm{real}$. Physical fields correspond to observation points (ϱ, t_r). a) Pulse profile at $\varrho = 0$ (t' is real at the source). b) Complex rays at (ϱ_1, t_r). Path 1 is a real space-time ray that describes the motion of the pulse peak with constant real frequency ω_0. Points A or B off the peak are reached by complex rays with frequencies $\omega(t'_2)$ and $\omega(t'_3)$, respectively. Each complex ray reaches only a single physical observation point. c) Real-space trajectories of points on the profile of a). Path 1 is a real space-time ray with constant real frequency ω_0. The other paths are traces of real-space intersections of complex rays. Since different points on these paths are reached by different complex rays, the frequency varies along each path

(1.156), where t' denotes the time variable at the source (Fig. 1.12a). The source frequency $\omega(t')$ is obtained from the negative time derivative of the (complex) source phase

$$\omega(t') = \omega_0 - (i/2\alpha^2)t' . \tag{1.163}$$

From (1.82) a wave packet with frequency $\omega(t')$ moves on a space-time trajectory defined by

$$\varrho/(t - t') = v_g[\omega(t')] . \tag{1.164}$$

Since $\omega(t')$ is complex for real t', see (1.163), $v_g[\omega(t')]$ is likewise complex, and (1.164) cannot then be satisfied for real (ϱ, t). Thus, to have (1.164) apply at real observation points (ϱ, t), it is necessary to have t' complex, and hence t complex because t' is the time variable at the source. Comparing (1.163), (1.164) and (1.159), one observes that $\omega(t')$ and ω_s are identical when t', the complex-ray parameter, is eliminated. On a complex ray identified by t', the frequency $\omega(t')$ remains constant. At the spatial or temporal peak of the pulse, $\omega = \omega_0$, whence $t' = 0$ and the ray trajectory in (1.162) is real. For space-time points not on the peak trajectory, ω and t' are complex and the ray configuration is as shown in Fig. 1.12b. Since each complex ray pierces the physical (ϱ, t_r) space only at a single point, the complex ray interpretation makes it evident that each observation point off the peak trajectory has its own frequency. From a plot of the real-space intersections of the complex rays (Fig. 1.12c) and the corresponding ray frequencies, one may ascertain the complete space-time evolution of the exponentially dependent portion of the pulse field in (1.84) (pulse spreading, frequency change, etc.). Evidently, tracking of wave packets[20] along complex rays is considerably more involved than tracking along real rays. Various procedures are presently being studied [1.35–37]. Graphical techniques like those in Fig. 1.5 can be employed but they now require plots of the real and complex parts of $k(\omega)$ when ω is complex.

1.6.5 Pulses in Dissipative Media—Group Velocity

The phase in (1.158) is complex also when the medium is lossy, i.e., when the propagation function $k(\omega)$ is complex for real ω. The saddle point frequency ω_s is then likewise complex and the general considerations in Subsection 1.6.4 apply here. In particular, when the Gaussian pulse in (1.156) excites a lossy medium, (1.159) and (1.160) remain valid. However, $v_g(\omega)$ is now complex for real ω. The real frequency $\bar{\omega}_s(\varrho, t_r)$ at the temporal maximum of the exponential envelope is given by

$$\bar{\omega}_s = \omega_0 - (\varrho/2\alpha^2)(\mathrm{Im}\{dk/d\omega\})_{\bar{\omega}_s}, \tag{1.165}$$

and the equation of motion of the temporal maximum by

$$\varrho/t_r = [\mathrm{Re}\{dk/d\omega\}]_{\bar{\omega}_s}^{-1}. \tag{1.166}$$

---

[20] The designation of "wave packet" applies here not to the Gaussian pulse as a whole but more generally to any small frequency interval within the pulse spectrum.

Since the frequency at the spatial peak does not remain constant, ϱ/t_r in (1.166) is not constant whence the spatial peak trajectory is curved. Referring to (1.163) and (1.165), one observes that the complex ray parameter t' is imaginary for rays characterizing observation points along the temporal peak path. Thus, (1.166) tracks the real-space intersections of complex rays. The resulting picture is as shown in Fig. 1.12c, except that even trajectory 1 is now replaced by a curved path descriptive of the real-space intersections of rays originating on the t_i axis. Moreover, it is found that the trajectories of the spatial and temporal peaks are no longer coincident, and the frequency variation along the spatial peak path differs from that along the temporal peak path [1.35].

The concept of group velocity $v_g = (dk/d\omega)^{-1}$ for describing the energy propagation characteristics of a wave packet with small frequency spread becomes obscured in a lossy medium since v_g is now complex. Furthermore, different propagation speeds may be associated with different features of the pulse (for example, the spatial and temporal maxima). Although attempts have been made to extend the notion of group velocity to wave packets in dissipative media by examining such quantities as $[\text{Re}\{dk/d\omega\}]^{-1}$, $\text{Re}\{d\omega/dk\}$, and other variants, there is little justification for choosing one definition over another. The only unambiguous definition is to retain $(dk/d\omega)^{-1}$ as a complex quantity without assigning to it the association with energy transport that is possible in lossless media. For weak dissipation or wide pulses, but for limited propagation lengths, the various definitions are approximately equivalent. However, under more general conditions, one must refer to the field solution in (1.84) to ascertain the propagation properties of the signal.

The input signal in (1.156) is non-causal since $-\infty < t < \infty$ in the source function. A causal solution initiated at time t_1 is obtained on multiplying $h(t)$ in (1.156) by the step function $U(t-t_1)$. The transient field at (ϱ, t) is then obtained more conveniently by integrating the product of the Green's function $G(\varrho, t, t')$ and the source function $h(t')$ from $t' = t_1$ to $t' = \infty$, see (1.25). In the resulting asymptotic evaluation, it is found that a saddle point at $t' = t'_s$ and the endpoint at $t' = t_1$ may contribute (for an evaluation of the endpoint contribution, see (1.167)). t'_s is defined as the solution of (1.163) (see also Fig. 1.12). One finds, qualitatively, that the saddle point, which furnishes the same solution as in the non-causal case, contributes only when $\text{Re}\{t'_s\} > t_1$, while the endpoint contributes for all $t' > t_1$. The absence of the saddle point contribution implies that no signal is observed for $t < t_1$. Since the endpoint contribution is proportional to the value of the integrand at t_1, see (1.167), its effect is small when the pulse is initiated early enough so that $h(t_1) \ll 1$.

1.6.6 The Utility of Ray Diagrams and Some Analogies between Diffraction Phenomena under Time-Harmonic and Time-Dependent Conditions

As the trajectories traversed by wave packets, space-time rays provide some fundamental insight into the propagation properties of transient signals. Many of the more familar characteristics of rays and the associated wavefront or high-frequency time-harmonic local plane wave fields in non-dispersive media have their counterpart in the dispersive transient regime. Some illustrations are given below. For the time-harmonic problem, we shall deal with two space dimensions (y, z); for the time-dependent dispersive case, the corresponding coordinates are (ct, z).

Focusing

When a ray system converges (i.e., the ray tube cross section decreases and the wavefront is concave), the local plane wave field amplitude increases up to the caustic or focal region and then decreases. A converging ray system in the time-harmonic field may be established by an initial phase distribution $S(y, 0)$ in an aperture plane $z = 0$ such that the spatial wavenumbers $\partial S(y, 0)/\partial y$ along y decrease or increase as y moves in the positive or negative direction away from a central value. This is schematized trivially in the last sketch of Fig. 1.13a, where the rays are drawn as normals to the refractive index surface defined in a reference medium with $n = 1$ by $(\partial S/\partial y)^2 + (\partial S/\partial z)^2 = 1$. When all rays do not pass through a single point $(\bar{y}, \bar{z})$, the focus, one has imperfect focusing as in the first sketch of Fig. 1.13a, with the resulting caustic and an evanescent field on the dark side.

In a lossless dispersive medium, a converging set of space-time rays may be produced by a frequency modulated input pulse, with the modulation so chosen that, for perfect focusing, all rays pass through a prescribed point $(c\bar{t}, \bar{z})$; the rays must then satisfy (1.164) with (ϱ, t) replaced by $(\bar{z}, \bar{t})$. From the considerations in Fig. 1.5, the desired frequency profile $\omega(t')$ in the input plane $z = 0$ is obtained by searching on the (ω, k) dispersion surface for those points with normal directions parallel to the corresponding rays. Space-time focusing corresponds to pulse compression [1.21]. When the focusing is imperfect, the ray system is bounded by space-time caustics, with exponentially small (evanescent) fields on the dark side (see Fig. 1.13b).

When the dispersive medium is lossy, the space-time rays are complex and caustics or foci generally also lie in complex coordinate space. This implies that the corresponding compression of the physical pulse

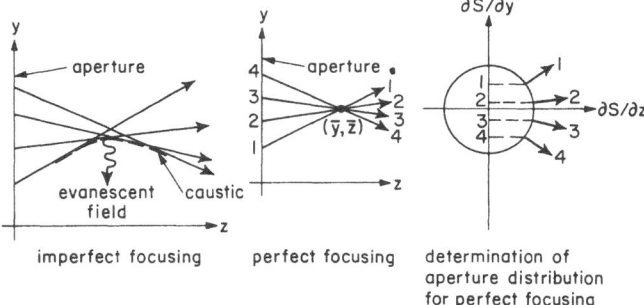

(a) phased aperture distribution

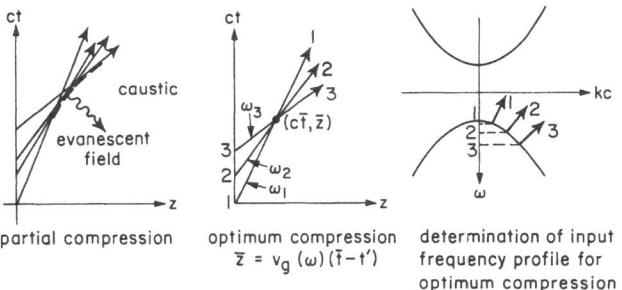

(b) frequency modulated input

Fig. 1.13a and b. Focusing shaped by initial conditions

field in real space is diminished. It is of interest to observe, however, that initial conditions may be defined so as to keep the focused ray system $\bar{z} = v_g(\omega)(\bar{t} - t')$ entirely in real space, i.e., $\bar{z}, \bar{t}, t', v_g$ real [1.35]. Since v_g in a lossy medium is complex for real ω, it follows that the input frequency profile $\omega(t')$ must be complex to admit real v_g. Thus, the input pulse for optimum compression is both amplitude and frequency modulated so as to equalize, for the various spectral components, the transmission losses and phase delays between the input plane and the compression point.

An originally diverging ray system in a non-dispersive environment may be made to converge by passing it through an isotropic lens with a curved contour. Alternatively, a diverging wavefront may be changed into a converging wavefront by refraction at a plane interface if the refractive index surfaces for the two media have opposite curvature (this generally implies anisotropy). The ray matching condition at the interface $z=0$ requires continuity of the tangential wavenumber $\partial S/\partial y$. Thus, the converging ray system in medium 2 may be obtained from the given ray system in medium 1 by the construction in Fig. 1.14a.

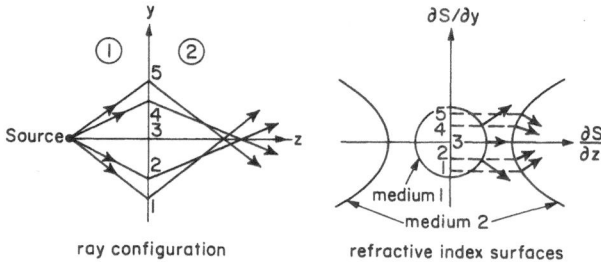

ray configuration   refractive index surfaces

(a) interface between an isotropic and anisotropic medium

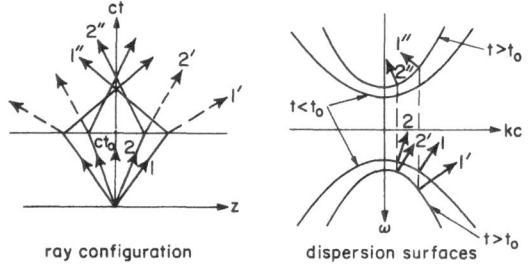

ray configuration   dispersion surfaces

(b) sudden temporal change in medium properties

Fig. 1.14a and b. Focusing by abrupt changes in the medium properties

For perfect focusing, one may employ the construction in reverse to determine the input ray system that yields a refracted ray system passing through a single point.

The same effect may be exploited for pulses in a dispersive environment if the medium properties are changed suddenly (and uniformly throughout the medium) at some time t_0 [1.38]. Since the wavenumber k, but not the frequency ω, in a wave packet is conserved during such a change, this process generates reflected and transmitted wave packets for $t>t_0$. The reflected wave packets (double primed in Fig. 1.14b) are described by a converging ray system and are therefore compressed; this behavior is due to the opposite curvatures of the relevant portions of the dispersion surfaces.

Diffraction Due to Truncation

When the time-harmonic focused aperture distribution in Fig. 1.13a is truncated at $y = \pm d$, corresponding to rays with propagation angles $\theta_1 < \theta < \theta_2$, the truncation edges act like localized sources of diffracted rays which propagate in all directions as in Fig. 1.15a. This may be

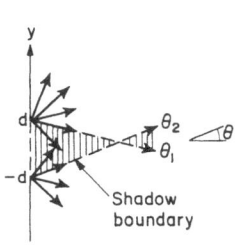

Finite focused aperture:
$-d < y < d\ (\theta_1 < \theta < \theta_2)$
(diffracted rays have
all values of θ)

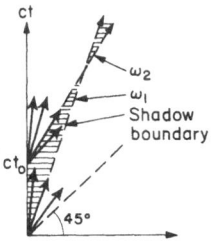

Square pulse envelope on FM
signal: $0 < t' < t_0\ (\omega_1 < \omega < \omega_2)$
(diffracted rays have all values of ω)

(a) Truncation in physical space. Diffracted fields
    originate at fixed coordinate locations

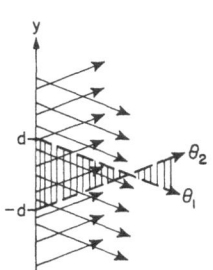

Finite angular spectrum:
$\theta_1 < \theta < \theta_2\ (-d < y < d)$
(diffracted fields emanate
from entire aperture plane)

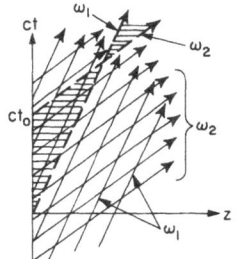

Finite frequency spectrum:
$\omega_1 < \omega < \omega_2\ (0 < t' < t_0)$
(diffracted fields exist
for all time)

(b) Truncated spectrum. Diffracted fields have line spectra

Fig. 1.15a and b. Diffraction due to end effects. Region traversed by main beam or signal rays is shown shaded

understood by taking the asymptotic form of the time-harmonic free-space Green's function $G_\omega(y, y'; z, 0)$, multiplying it by the aperture phase distribution $\exp[i(\omega/c)S(y',0)]$, and integrating over y' from $y'=-d$ to $y'=+d$. The stationary points in the integration interval provide the field in the shaded main beam region of Fig. 1.15a while the endpoints $y'=\pm d$ of the integration interval yield the contribution [1.17]

$$\int_{}^{d} dy'\, A(y')\exp[i(\omega/c)B(y')] \tag{1.167}$$
$$\sim A(d)\exp[i(\omega/c)B(d)]\,[i(\omega/c)(dB/dy')_d]^{-1},$$

and similarly for $y'=-d$. Since $A(d)\exp[i(\omega/c)B(d)]=G_\omega(y,d;z,0)\cdot$ $\exp[i(\omega/c)S(d,0)]$ behaves like a source located at $y=d$, the above conclusion follows. The diffraction due to truncation of the aperture is of the same form as the edge diffraction when the incident field illuminates a slit aperture a width $2d$ in an infinite plane.

Analogous phenomena in a dispersive medium arise when the frequency modulated input signal in Fig. 1.13b has a square envelope that limits its duration to the time interval $0<t'<t_0$. The frequency spectrum corresponding to this time interval is taken to be $\omega_1<\omega<\omega_2$. Referring to (1.167), with y' replaced by t', G_ω by the impulse-excited field G, and the aperture phase by the phase of the FM input, one concludes that the truncation of the input signal generates diffracted fields that appear to originate at $t'=0$ and $t'=d$; these suddenly initiated excitations have a spectrum that encompasses all frequencies (Fig. 1.15a). By analogy with the time-harmonic problem, these fields describe edge diffraction in time.

The focused aperture distribution in Fig. 1.15a may be truncated in an alternative manner by limiting its angular spectrum to lie within the interval $\theta_1<\theta<\theta_2$ as in Fig. 1.15b. For the same focusing condition as in Fig. 1.15a, this spectral range of the input falls into the spatial range $-d<y<d$ in the aperture plane. The truncation of the spectrum implies a filter function V such that the integration in the representation of (1.74) extends only from $\xi=\xi_1$ to $\xi=\xi_2$. Stationary points in the integration interval provide the same main beam region as in Fig. 1.15a; however, the endpoint contributions now arise from ξ_1 and ξ_2 which describe waves with angular directions θ_1 and θ_2, respectively. When the appropriately modified formula (1.167) is applied to endpoint contributions ξ_1 and ξ_2, one finds that these local plane waves exist over the entire aperture plane, as indicated in Fig. 1.15b, although their strength decreases away from the main beam region.

In the corresponding pulse-excited dispersive problem, a frequency filter restricts the spectrum to the interval $\omega_1<\omega<\omega_2$; this frequency range of the input falls within the time interval $0<t'<t_0$. The effect of truncation of the Fourier inversion integral (1.80) is now to provide

diffracted wave packets at the constant frequencies ω_1 and ω_2 which exist for all time (but with an amplitude that decreases away from the main signal region) (Fig. 1.15b).

The diffraction effects discussed above are de-emphasized when the main beam or main signal amplitude tapers off toward the truncation edges since the diffracted fields are proportional to the strength of the edge illumination.

Gaussian Beams and Wave Packets

When a time-harmonic aperture distribution has a Gaussian amplitude dependence $\exp(-kCy^2)$, C being a positive constant, the saddle points ξ_s in the integral (1.74) are complex. Accordingly, the field in (1.79) is described in terms of local plane waves with complex phase that move along complex ray trajectories. The observable field is given by the real-space intersections of the complex rays and defines a Gaussian beam [1.39]. The interpretation of the diagram in Fig. 1.16 is directly analogous to that for the Gaussian wave packet in Fig. 1.12.

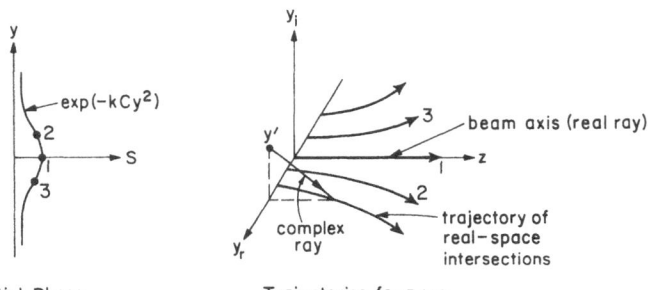

Fig. 1.16. Gaussian beam. For Gaussian wavepacket, see Fig. 1.12

Acknowledgement

The author expresses his appreciation to the John Simon Guggenheim Memorial Foundation for granting him a Fellowship during the academic year 1973–74, part of which was devoted to the preparation of this chapter. The author also acknowledges with thanks the aid of Mrs. Anna Mae Cuomo for her careful typing and editing of the manuscript.

References

1.1   L. B. Felsen, N. Marcuvitz: *Radiation and Scattering of Waves* (Prentice Hall, Inc., Englewood Cliffs, N. J. 1973)

1.2   Reference 1.1, Sect. 1.5b

1.3   M. KLINE, I. W. KAY: *Electromagnetic Theory and Geometrical Optics* (Interscience Publishers, Wiley and Sons, New York 1965) Sect. 1.3

1.4   Reference 1.1, Sects. 1.1b and 1.1c

1.5   YU. A. KRAVTSOV, L. A. OSTROVSKY, N. S. STEPANOV: Proc. IEEE **62**, 1492 (1974)

1.6   V. L. GINZBURG: *The Propagation of Electromagnetic Waves in Plasmas* (Pergamon Press, New York 1964)

1.7   C. T. TAI: *Dyadic Green's Functions in Electromagnetic Theory* (Intext Educational Publishers, Scranton, Pa. 1971)

1.8   J. A. STRATTON: *Electromagnetic Theory* (McGraw-Hill Book Co., New York 1941) Chapt. 1

1.9   J. B. KELLER, A. BLANK: Comm. Pure Appl. Math. **4**, 75 (1951)

1.10 Reference 1.1, Chapt. 3

1.11 Reference 1.1, Sect. 1.3

1.12 L. A. WEINSTEIN: *Open Resonators and Open Waveguides* (Golem Press, Boulder, Colo. 1969) Chapts. 10 and 11

1.13 L. B. FELSEN: Quart. Appl. Math. **23**, 151 (1965)

1.14 J. B. KELLER: J. Opt. Soc. Am. **52**, 116 (1962)

1.15 G. A. DESCHAMPS: Proc. IEEE **60**, 1022 (1972)

1.16 W. MAGNUS, F. OBERHETTINGER: *Formulas and Theorems for the Functions of Mathematical Physics* (Chelsea Publishing Co., New York 1964) Chapt. 3

1.17 Reference 1.1, Chapt. 4 and Sect. 1.6

1.18 L. B. FELSEN: IEEE Trans. AP-17, 191 (1969)

1.19 L. B. FELSEN, G. DESCHAMPS (guest editors): *Rays and Beams*, Special Issue, Proc. IEE **62** (Nov. 1974)

1.20 Reference 1.1, Sects. 1.6 and 1.7

1.21 L. B. FELSEN: IEEE Trans. AP-**19**, 424 (1971)

1.22 Reference 1.1, Sects. 6.2 and 6.3

1.23 A. ERDELYI, W. MAGNUS, F. OBERHETTINGER, F. G. TRICOMI: *Tables of Integral Transforms* (Mc Graw Hill Book Co., N. J. 1954) Vol. I, 26; Vol. II, 3, formula (35)

1.24 J. H. THOMPSON: SIAM J. Appl. Math. **22**, 300 (1972)

1.25 L. B. FELSEN: SIAM J. Appl. Math. **26**, 306 (1974)

1.26 Reference 1.1, Sects. 3.4 and 6.8

1.27 F. OBERHETTINGER, R. F. DRESSLER: J. Appl. Math. Phys. (ZAMP) **22**, 937 (1971)

1.28 Reference 1.16, Chapts. 4 and 8

1.29 J. J. BOWMAN, T. B. A. SENIOR, P. L. E. USLENGHI: *Electromagnetic and Acoustic Scattering by Simple Shapes* (North-Holland Publ. Co., Amsterdam 1969) Chapt. 18

1.30 D. S. JONES: *The Theory of Electromagnetism* (The MacMillan Co., New York 1964) Chapt. 10

1.31 E. OTT, J. SHMOYS: Canad. J. Phys. **46**, 1059 (1968)

1.32 D. G. DUDLEY, T. M. PAPAZOGLOU, R. C. WHITE: J. Appl. Phys. **45**, 1171 (1974)

1.33 D. A. HILL: J. Appl. Phys. **43**, 898 (1972)

1.34 J. R. WAIT: Radio Sci. 69D, 1387 (1965)

1.35 K. CONNOR, L. B. FELSEN: Proc. IEEE **62**, 1586 (1974)

1.36 J. A. BENNETT: Proc. IEEE **62**, 1577 (1974)

1.37 K. SUCHY: Proc. IEEE **62**, 1571 (1974)

1.38 L. B. FELSEN, G. M. WHITMAN: IEEE Trans. AP-**18**, 242 (1970)

1.39 S. CHOUDHARY, L. B. FELSEN: Proc. IEEE **62**, 1530 (1974)

1.40 R. G. KOUYOUMJIAN: The Geometrical Theory of Diffraction and Its Application; in *Topics in Applied Physics*, Vol. 3, edited by R. MITTRA (Springer Berlin, Heidelberg, New York 1975) pp. 165–215

2. Integral Equation Methods for Transient Scattering

R. MITTRA

With 33 Figures

Current developments in high resolution radar and EMP technology have created interest in the investigation of radiation and scattering of transient waveforms from conducting bodies of various shapes. The solution of such problems also finds application in target identification, inverse scattering, electromagnetic compatibility, and so on.

There are two independent techniques available for solving transient electromagnetic problems. The first of these involves the computation of the frequency-domain response of the structure, which is subsequently Fourier transformed to yield the desired time-domain response. Much has been written on the subject of frequency-domain solutions of electromagnetic scattering and radiation problems using, primarily, the integral equation technique as applied to time-harmonic signals. The interested reader has ready access to a vast array of literature including a recently published text [2.1] which covers this topic in considerable detail. In this work we will describe an alternative approach, which is based on a direct formulation of the integral equation in the time domain and its subsequent solution using a time-stepping rather than matrix inversion algorithm. After a brief survey of the historical development presented in Section 2.1, we go on to describe the derivation and solution of the general integral equation for the scattering problem in Subsection 2.2.1. The specialization to two- and one-dimensional cases, i.e., cylindrical and wire type radiators and scatterers, leads to special forms of integral equations that merit separate discussion, and will be covered in Subsections 2.2.2 and 2.2.3. Numerical aspects, which represent an important consideration in the construction of the solution of all of these equations will be discussed in Section 2.3 in considerable detail. Representative results derived through the use of time-domain integral equations of various kinds will be presented in Section 2.4. A comparative study of the time-domain vs. frequency-domain approaches will be given in Section 2.5 with the purpose of acquainting the reader with the relative merits and disadvantages of these two competing techniques.

In Section 2.6 we introduce a technique for representing the time-domain response of a system in the form of a series of complex exponentials. The basis for this type of expansion, referred to as SEM (singularity expansion method), is discussed in Chapter 3 on the basis of the frequency-domain formulation. Our objective here will be to show that the SEM expansion can also be derived directly from the time-domain results.

Finally, some discussion pertaining to the extensions and applications of time-domain methods will be presented in Sections 2.7 and 2.8. It is hoped that this chapter will be able to provide the reader with a good perspective of the scope and limitations of the time-domain approaches currently available for solving transient electromagnetic problems.

2.1 Brief Review of Historical Developments

The first attempts at deriving time-domain solutions to scattering problems date back several years. KENNAUGH and COSGRIFF [2.2] employed a physical-optics approximation to calculate backscatter impulse responses of a rectangular flat plate, sphere and spheroid. They improved their approximate first results by making these results satisfy the so-called moment conditions. KENNAUGH and MOFFATT [2.3] have summarized the subsequent work which was devoted to extending the application of physical and geometrical optics to transient response calculations. WEEKS [2.4] has also used the physical-optics approximation in conjunction with the reciprocity theorem to approximately calculate the impulse response of a number of simple scatterers. RHEINSTEIN [2.5] has described a technique based upon the Fourier transformation of the frequency domain solution in the form of the classical Mie series and has computed the short pulse response of a sphere.

The physical-optics approach is quite useful for computing the approximate impulse response of certain target shapes. However, this approach typically requires that the scatterer satisfy certain conditions, e.g., the characteristic dimensions of the scatterer must be large compared with the wavelength. On the other hand, the series type of solution converges rather slowly for large and even for moderately large body sizes. More importantly, the series solution can be obtained only for a rather restricted class of separable geometries. Finally, as will be shown in the following section, it is possible to derive the time-domain solutions directly rather than via Fourier transformation of the frequency-domain solution.

In 1968, BENNETT and WEEKS [2.6] introduced a novel approach for solving the time-domain integral equation and presented extensive numerical data derived from the use of this approach. They dealt with both two- and three-dimensional perfectly conducting scatterers and showed that a solution to the time-dependent integral equation can be constructed without resorting to a matrix solution, but using a time-stepping approach that is similar to that employed for solving initial value problems. The form of the equation used by BENNETT and WEEKS was the Magnetic Field Integral Equation [2.7] or MFIE. SAYRE and HARRINGTON [2.8] also used a time-stepping approach to derive the transient response of wire scatterers, but they employed the Electric Field Integral Equation [2.7], or EFIE, to solve this problem. Although basically similar, there are certain fundamental differences between the way the above two forms of the integral equations are handled. This will be evident from the discussion appearing in Sections 2.2 and 2.3.

Additional work on time-domain calculations for a wide range of structures has been reported in a number of papers by BENNETT and his co-workers [2.6, 9–11]. MILLER et al. [2.12] and POGGIO et al. [2.13] have investigated wire structures in some detail. MILLER has also written an excellent summary work [2.14] on the computational aspects of transient electromagnetics. In this chapter we will freely draw upon the material that has appeared in many of these publications and present brief descriptions of several forms of time-domain integral equations and their solutions.

2.2 Time-Domain Integral Equations

2.2.1 General Three-Dimensional Structures

The integral equation approach represents a convenient means for formulating electromagnetic boundary value problems. In this work we will make exclusive use of this approach for the problems we consider, although other methods [2.15], e.g., finite difference schemes, may also be applied to certain types of transient scattering problems. As mentioned in the last section there are two types of integral equations, viz., the MFIE and EFIE. The derivation of both of these equations is described below.

The Electric and Magnetic Field Integral Equations (EFIE and MFIE)

We begin with the time-dependent forms of Maxwell's equations

$$\nabla \times E(r,t) = -\frac{\partial}{\partial t} \mu_0 H(r,t) \qquad (2.1\,\text{a})$$

$$\nabla \times H(r,t) = \frac{\partial}{\partial t} \varepsilon_0 E(r,t) + J(r,t) \tag{2.1b}$$

$$\nabla \cdot E(r,t) = \varrho(r,t)/\varepsilon_0 \tag{2.1c}$$

$$\nabla \cdot H(r,t) = 0, \tag{2.1d}$$

where μ_0 and ε_0 are the permeability and permittivity of free space, respectively, r is the position vector, J is the current density and ϱ is the charge density. The last two quantities are related by the continuity equation

$$\nabla \cdot J + \frac{\partial}{\partial t} \varrho(r,t) = 0. \tag{2.2}$$

It is convenient at this stage to introduce the scalar and vector potentials ϕ and A as

$$H(r,t) = \frac{1}{\mu_0} \nabla \times A(r,t) \tag{2.3}$$

and

$$E(r,t) = -\nabla \phi(r,t) - \frac{\partial A}{\partial t}(r,t). \tag{2.4}$$

These potentials may be related to each other via the Lorentz gauge condition

$$\nabla \cdot A(r,t) + \mu_0 \varepsilon_0 \frac{\partial}{\partial t} \phi(r,t) = 0. \tag{2.5}$$

Using Maxwell's equations (2.1), (2.3) and (2.4), we can readily derive the wave equations satisfied by A and ϕ

$$\nabla^2 A(r,t) - \mu_0 \varepsilon_0 \frac{\partial^2}{\partial t^2} A(r,t) = -\mu_0 J(r,t) \tag{2.6}$$

$$\nabla^2 \phi(r,t) - \mu_0 \varepsilon_0 \frac{\partial^2}{\partial t^2} \phi(r,t) = -\varrho(r,t)/\varepsilon_0. \tag{2.7}$$

The solution to the above equations is readily constructed using an auxiliary equation for the Green's function which reads

$$\nabla^2 g - \mu_0 \varepsilon_0 \frac{\partial^2 g}{\partial t^2} = \delta(\mathbf{r} - \mathbf{r}', t - t') . \tag{2.8a}$$

The solution of (2.8a) is given by

$$g(\mathbf{r}, t, \mathbf{r}', t') = -\frac{1}{4\pi R} \delta(t - t' - R/c) \tag{2.8b}$$

where

$$R = |\mathbf{R}| = |\mathbf{r} - \mathbf{r}'|$$

$$c = 1/(\mu_0 \varepsilon_0)^{1/2} = \text{velocity of light}$$

$$\delta(x) = \text{delta function} .$$

The appropriate solutions are

$$\phi(\mathbf{r}, t) = \frac{1}{4\pi\varepsilon_0} \int_V \frac{\varrho(\mathbf{r}', t - R/c)}{R} dv' \tag{2.9a}$$

$$A(\mathbf{r}, t) = \frac{\mu_0}{4\pi} \int_V \frac{J(\mathbf{r}', t - R/c)}{R} dv' , \tag{2.9b}$$

where the volume V is the support of the current and charge distributions J and ϱ. Inserting these expressions for ϕ and A into (2.3) and (2.4) leads to the sought-for integral representation for the time-dependent electric and magnetic fields produced by the sources J and ϱ. These expressions are

$$E(\mathbf{r}, t) = -\frac{1}{4\pi\varepsilon_0} \int_V \frac{1}{R^2} \nabla R(\mathbf{r}', \tau) \varrho(\mathbf{r}', \tau) dv'$$

$$+ \frac{1}{4\pi\varepsilon_0} \int_V \frac{1}{R} \nabla R \frac{\partial}{\partial \tau} \varrho(\mathbf{r}', \tau) dv' \tag{2.10a}$$

$$+ \frac{\mu_0}{4\pi} \int_V \frac{1}{R} \frac{\partial}{\partial \tau} J(\mathbf{r}', \tau) dv'$$

and

$$H(\mathbf{r}, t) = \frac{1}{4\pi} \int_V \left[\frac{1}{R^2} J(\mathbf{r}', \tau) + \frac{1}{Rc} \frac{\partial}{\partial \tau} J(\mathbf{r}, \tau) \right] \frac{\mathbf{R}}{R} dv' \tag{2.10b}$$

with

$$\tau = t - R/c = \text{retarded time}$$

and

$$\frac{\partial}{\partial \tau} f(\tau) = \frac{\partial f(t)}{\partial t}\bigg|_{t=\tau}.$$

In this chapter we will be concerned with perfectly conducting bodies, which support induced currents and charges only on their surfaces. For such surface source distributions, the E- and H-field representations take the form

$$E(r,t) = -\frac{1}{4\pi} \int_S \frac{1}{R} \left\{ \frac{1}{\varepsilon_0 R} \nabla R \sigma(r',\tau) \right.$$

$$\left. + \frac{1}{\varepsilon_0} \frac{\partial}{\partial \tau} \sigma(r',\tau) \nabla R + \mu_0 \frac{\partial}{\partial \tau} J_s(r',\tau) \right\} ds' \tag{2.11a}$$

and

$$H(r,t) = \frac{1}{4\pi} \int_S \left\{ \frac{J_s(r',\tau)}{R} + \frac{1}{c} \frac{\partial}{\partial \tau} J_s(r,\tau) \right\} \times \frac{R}{R} ds'. \tag{2.11b}$$

The Electric Field Integral Equation, EFIE, for the scattering problem involving a perfect conductor may now be readily derived from

$$\hat{n} \times E_{\text{tot}} = \hat{n} \times (E + E^{\text{inc}}) = 0 \tag{2.12}$$

on the surface on the conductor, i.e., $r \in S$, S being the surface of the conductor and $\hat{n}$ the unit outward normal to S. As a result, we obtain the following equation

$$\hat{n} \times E^{\text{inc}}(r,t) = \frac{\hat{n}}{2\pi} \times \int_S \left[\frac{\mu_0}{R} \frac{\partial}{\partial \tau} J_s(r',\tau) \right.$$

$$\left. - \frac{\sigma(r',\tau)}{\varepsilon_0} \frac{R}{R^3} - \frac{1}{\varepsilon_0} \frac{\partial}{\partial \tau} \sigma(r',\tau) \cdot \frac{R}{cR^2} \right] ds', \tag{2.13}$$

where we have explicitly written out ∇R for convenience of manipulation later. In addition, one may note that the singularities of the kernel in the integral representation, obtained by letting $r \to r'$ in (2.11a), require that they be interpreted properly [2.7, 16]. When this is done, the integral is found to be the Principal Value (P.V.) type, as indicated in (2.13). The P.V. integral is interpreted as

$$\int_S = \lim_{\Sigma \to 0} \int_{S-\Sigma} .$$

The derivation of the Magnetic Field Integral Equation, or MFIE, is done somewhat differently. The appropriate boundary condition to be imposed for deriving this equation is on the normal component of the magnetic field. We let

$$\hat{n} \times H_{tot} = J_s \tag{2.14}$$

which enforces the boundary condition (2.12) on the tangential electric field only in an indirect manner.

Applying the condition (2.14) to the representation (2.11b) yields the time-dependent MFIE

$$J_s(r,t) = 2\hat{n} \times H^{inc} + \frac{1}{2\pi} \hat{n} \times \int_S \left\{ \frac{1}{c} \frac{\partial}{\partial \tau} J_s(r',\tau) + \frac{J_s(r',\tau)}{R} \right\} \times \frac{R}{R^2} ds' \tag{2.15}$$

where, once again, the integral involved is of P.V. type. As will be seen shortly, the special forms of the time-domain integral equations (2.13) and (2.15) play a fundamental role in enabling us to construct their solution using a time-stepping procedure. For comparison we reproduce here the corresponding integral equations in the frequency domain (see [2.7] for extensive discussion of these equations)

$$\hat{n} \times E^{inc}(r) = \frac{1}{2\pi i \omega \varepsilon_0} \hat{n} \times \int \left\{ -\omega^2 \mu_0 \varepsilon_0 J_s(r') \phi + [\nabla' \cdot J_s(r')] \nabla' \phi \right\} ds' \tag{2.16}$$

$$J_s(r) = 2\hat{n} \times H^{inc}(r) + \frac{\hat{n}}{2\pi} \times \int_S J_s(r') \times \nabla' \phi \, ds' \tag{2.17}$$

with $\phi = \exp(ikR)/R$

and $k = \omega\sqrt{\mu_0 \varepsilon_0}$.

In contrast with (2.13) and (2.15), these equations in the frequency domain are handled numerically [2.7] by matrix inversion, rather than by initial value techniques that are applicable to their time-domain counterparts.

Solution of Integral Equations for Three-Dimensional Scatterers

We now proceed to outline the principles that form the basis for constructing the solution to the integral equations which we have just derived. The equation which is most convenient for three-dimensional scatterers is the MFIE given in (2.15). There are two unique features of this integral equation for the induced surface current J_s. First, it is evident that this equation is of the second kind. Second, for a given time t, the unknown J_s *inside* the integral has an argument $\tau = t - R/c$. Since the P.V. integral excludes the point $R = 0$, τ is always less than t. In view of the above, one may regard (2.15) as an expression for the unknown current $J_s(t)$, at any given time t, in terms $2\hat{n} \times \bar{H}_{inc}$ containing the known incident term, and an integral which is also known from the past history of the current J_s.

This point may be further illustrated by reference to Fig. 2.1 which shows an incident signal impinging on a solid-surface scatterer. Consider

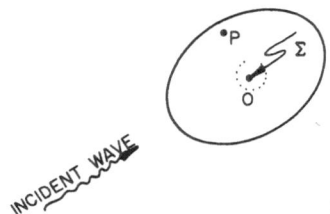

Fig. 2.1. General three-dimensional scattering problem

the problem of computing the induced surface current density at the latest observation time at a point 0 located on the scatterer.

According to (2.15) the surface current density at the point 0 for a given time is given by $2\hat{n} \times H^{inc}$ at that point plus the integral appearing in the right-hand side of (2.15). This integral may be broken up into two parts. The first of these represents the contribution of the magnetic field generated by the induced currents everywhere excepting those currents that exist in Σ, which is a small patch centered around 0. The second is the contribution of the P.V. integral at 0 due to the induced currents in the self-patch Σ itself. However, it can be shown that the second contribution is identically zero if Σ is a flat surface, and is negligibly small if the principal radii of curvatures of Σ are large compared to the wavelength of the highest spectral component of the incident wave. Thus, if the self-patch P.V. integral is neglected, i.e., if only the first of these two contributions is retained, then all of the induced surface current contributions at 0, due to currents elsewhere, e.g., the neighborhood of points such as P, that are separated from the point 0 by a finite distance $\overline{OP}$, will be retarded in time by $|\overline{OP}|/c$. Consequently, only the induced currents that existed there *before* the time $t - \overline{OP}/c$ can contribute to the fields at point 0 *at time* t, and the fields radiated by all the currents that have been induced at points such as P since that time will not have reached 0 because of time retardation. It follows that if all of the past history of the induced current at P is available up to the time $t - |\overline{OP}|/c$ we can compute the contribution of these currents at the point of observation 0 at the observation time t. This procedure forms the basis of an iterative technique for constructing the solution of (2.15). The solution is started at the time $t = 0$, invoking causality which guarantees that all past currents and their derivatives are identically zero for $t < 0$. The solution is then constructed by gradually building it up for $t > 0$, by marching on time in a manner which is very similar to that employed for solving an initial value problem using a time-stepping procedure. It should be pointed out that in making the above statement we have tacitly assumed that not only $J_s(\tau)$ but $\partial J_s(\tau)/\partial \tau$ also is derivable at the observation time t from the past history of J_s.

The method for solving the EFIE, given in (2.13), is somewhat more involved, but the same general procedure can still be employed. One major difference between the EFIE and MFIE is that for the former the P.V. integral evaluated for the self-patch is not zero, as it is for the latter, even when the patch size is small. This could have been predicted easily, since, for any given observation time, the integral in the r.h.s. of (2.13), which generates the scattered electric field, would contain only *retarded time* in its argument if the self-patch were excluded. However, the only term in the l.h.s. of (2.13) is $\hat{n} \times E^{inc}$, which is clearly nonzero at the observation time t. This would imply that the two sides of (2.13) cannot equal each other unless there is a flaw in the argument presented above. It turns out that this discrepancy is readily resolved by studying the self-patch integral carefully. Upon doing this one finds that the self-patch, P.V. integral in (2.13) contributes a non-zero term of the type $a_{nn}J_s(r_{0n}, t)$, where a_{nn} can be derived from the evaluation of the self-patch integral and r_{0n} is the point of observation. Hence, J_s can be simply calculated from

$$J_s(r_{0n}, t) = (1/a_{nn}) \left[\hat{n} \times E^{inc}(r_{0n}, t) + \oint_s \begin{Bmatrix} \text{terms involving} \\ \text{retarded time } t < 0 \end{Bmatrix} \right] ds'$$

(2.18)

where $\oint$ indicates the integral from which the self-patch has been excluded. One can now apply the same type of time-stepping procedure as employed in connection with the MFIE to extract the solution of the EFIE in the time-domain.

It is useful to point out here that several variants of this basic procedure are possible by working with different versions of the E equation that are mathematically equivalent but may have profoundly different numerical interpretation. This point is further illustrated with two examples in Subsections 2.3.2 and 2.3.3.

Typically, for solid surface structures, the H integral equation is preferable because its kernel is less singular than the E equation. As a consequence, less sophisticated expansion functions may be employed for representing the unknown current. On the other hand, the MFIE is entirely unsuitable for flat surfaces [2.17], or wire structures as this equation becomes unstable for electrically thin structures.

2.2.2 Time-Domain Integral Equation for Two-Dimensional Surfaces

Let us now consider the integral equation for cylindrical structures that are independent along one of the coordinate axes, e.g., the z-axis [see Fig. 2.2.] If the incident wave is also independent of the z-coordinate,

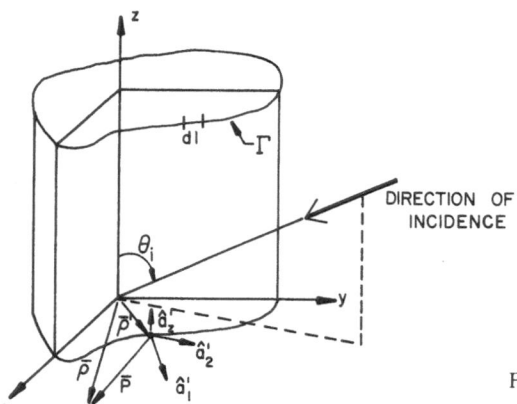

Fig. 2.2. Geometry of cylindrical scattering problem

one can follow the usual route of integrating out the z-variation from the three-dimensional integral equations (2.13) and (2.15). However, because the variable of this integration, viz., z', appears not only in the kernels of these equations but also in the arguments of the unknown functions by way of $\tau = t - |r - r'|/c$, this integration is not easily performed. It is more convenient, instead, to revert to (2.6) and (2.8) which define the vector potential A and the Green's function g. For convenience we reproduce these equations below

$$\nabla^2 A - \mu_0 \varepsilon_0 \frac{\partial^2}{\partial t^2} A = -\mu_0 J(r,t) \tag{2.6}$$

$$g(r,t; r',t') = -\frac{1}{4\pi R} \delta(t - t' - R/c). \tag{2.8}$$

We may solve (2.6) using (2.8) and write

$$
\begin{aligned}
A &= \frac{1}{4\pi\mu_0} \int dv' dt' \frac{J(\varrho',t)\delta(t - t' - R/c)}{R} \\
&= \frac{1}{4\pi\mu_0} \int_\Gamma \int_{-\infty}^{0} \int_{-\infty}^{\infty} J(\varrho',t')\delta \frac{(t - t' - R/c)}{R} dz'\, dt'\, dl',
\end{aligned}
\tag{2.19}
$$

where l is the integration variable along the contour Γ of the cylindrical cross section (see Fig. 2.2).

Since the unknown current $J(\varrho',t')$ is independent of z', this equation is convenient for integrating out the z'-variable, as long as this integration is performed before the t' integration. Integrating with respect to the z' variable, we obtain from (2.19) (see also (1.138) with $\omega_p=0$, for the two-dimensional Green's function)

$$A(\varrho,t) = \frac{c}{2\pi\mu_0} \int_\Gamma dl \int_{-\infty}^{t - P/c} \frac{J_s(\varrho',t')}{\sqrt{c^2(t-t')^2 - P^2}} dt' \tag{2.20}$$

where $P=|\varrho - \varrho'|$ and Γ is the contour defining the cylindrical surface. We may now express the scattered H field as $\nabla \times A$; thus, the total tangential H field $(=J_s)$ becomes

$$
\begin{aligned}
J_s = 2\hat{n} \times H^{inc}(\varrho,t) + \frac{c\hat{n}}{\pi} \int_\Gamma dl(\hat{n} \cdot \hat{n}') \int_{-\infty}^{t - P/c} \left[\frac{dt'}{\sqrt{c^2(t-t')^2 - P^2}} \right] \\
\times \left[\frac{J_s}{c(t-t') + P} + \frac{1}{c} \frac{\partial}{\partial t'} J_s \right]
\end{aligned}
\tag{2.21}
$$

where $\hat{n}$ and $\hat{n}'$ are unit outward normals in the (P,Q) and (P',Q') systems, respectively. One notes from the above integral equation for the two-dimensional problems that in contrast to the general three-dimensional problem the two-dimensional vector integral equation (2.21) is separable into two, uncoupled scalar equations. This, of course, was to be expected from the well-known result that the fields in the cylindrical structure are separable into TE and TM components. Another point to note in (2.21) is that in common with the three-dimensional equation (2.15), the integral in the r.h.s. of (2.21) contains J_s with retarded time in its argument[1]; hence, the contribution of the integral is completely known at a time t in terms of the quantities that have already been computed previously. Thus, once again, it is possible to construct a solution to (2.21) simply by marching on time.

There is, however, one important difference here from the three-dimensional case that is worth noting. The integral in the r.h.s. of (2.21) contains an integration with respect to t' which was non-existent in (2.15). This time integration can be carried out in the usual manner, by resorting to numerical techniques.

2.2.3 The One-Dimensional Case

Having discussed the formulation of the general, three-dimensional time-domain integral equations and the specialization of the H integral equation to the two-dimensional case, we now turn to one-dimensional structures, viz., the wire problem. The geometry of the problem is shown in Fig. 2.3. The derivation of the integral equation follows along much the same lines as presented earlier. Under the so-called thin-wire approximation, we assume that the flow of the induced current is restricted along the s-direction[2], which defines the wire (see Fig. 2.3). It is also assumed that there is no azimuthal variation of the current so that the unknown current I is a function of s only. Under these conditions the E integral equation may be written as

$$
\hat{s} \cdot E^{\mathrm{inc}}(r,t) = \frac{\mu_0}{4\pi} \int \left[\frac{\hat{s} \cdot \hat{s}'}{R} \frac{\partial}{\partial \tau} I(s', \tau) \right.
$$

$$
\left. + c \frac{\hat{s} \cdot R}{R^2} \frac{\partial}{\partial s'} I(s', \tau) + c^2 \frac{\hat{s} \cdot R}{R^3} \int_{-\infty}^{\tau} \frac{\partial}{\partial s'} I(s' \cdot t') dt' \right] ds',
$$

$$
r \in C + a \tag{2.22}
$$

---

[1] The reason this integral is not written as P.V. is that the kernel is never singular and, hence, there is no need to exclude any portion of the integral to define a P.V. integral.

[2] Here, s is not to be confused with the same symbol used in Subsect. 2.3.4, Sect. 2.6 and in other chapters for the Laplace transform (complex frequency) variable.

where $\hat{s}$ and $\hat{s}'$ are the unit tangent vectors, C is the wire contour, a is the wire radius and $\tau = t - |r - r'|/c$.

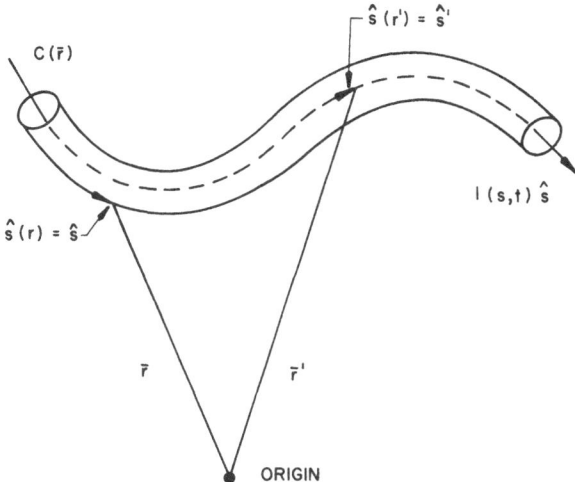

Fig. 2.3. Thin wire showing filamentary path and spatial relationships

Note that as in (2.21) the one-dimensional time-domain integral equation (2.22) contains the integral w.r.t. time, although for a different reason. The time-integration appears in the EFIE with which we are dealing when the expression for charge distribution, viz.,

$$q(s',\tau) = -\int_{\infty}^{\tau} \frac{\partial}{\partial s'} I(s',t')dt'$$

is inserted in the representation of the scattered E field.

The one-dimensional equation given in (2.22) can again be solved by a time-stepping procedure by first isolating the self-patch term in the same manner as discussed in connection with the solution of three-dimensional EFIE discussed in Subsection 2.2.1. The rest of the solution also follows along the same lines as discussed earlier in connection with the general three-dimensional case of EFIE. Finally, before closing this section it may be worth mentioning once again that the H integral equation becomes numerically unstable for electrically thin wires and hence is unsuitable.

2.3 Numerical Solution of Integral Equations

2.3.1 The Magnetic Field Integral Equation (MFIE)

From the numerical point of view one of the most important features of the MFIE equations (2.15) and (2.21) is that the kernels have a lower-order singularity as compared to the EFIE. This feature allows one to employ basis and testing functions (see [2.18] for definition of these functions) that are simpler than the corresponding ones for the electric field integral equation. One can choose, for instance,

$$J_s(r,t) = \sum_{L=1}^{N_s} \sum_{j=1}^{N_t} A_{ij}(r-r_i; t-t_j) V(r_i) U(t_j), \qquad (2.23)$$

where $V(r_i) = 1$ for r on the surface patch of area ΔS_i centered at r_i on the scatterer and 0 elsewhere,
$U(t_j) = 1$, for t in the time interval Δt centered at t_j,
$= 0$, elsewhere.

The unknowns A_{ij} are the vector weight coefficients for the space-time sampled values of the current J_s on the surface of the structure which is assumed to be subdivided into N_s patches centered at r_i (see Fig. 2.4). The total time is also assumed to be subdivided into N_T time-steps.

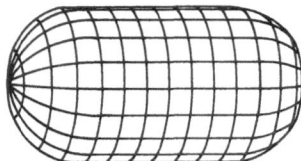

Fig. 2.4. Subsectionalization of a surface scatterer

The next step is to approximate the integral appearing in the MFIE which is reproduced here for the sake of convenience

$$J_s(r,t) = 2\hat{n} \times H^{inc}(r,t)$$

$$+ \frac{1}{2\pi} \hat{n} \times \int \left\{ \frac{1}{c} \frac{\partial}{\partial \tau} J_s(r',\tau) + J_s(r',\tau) \frac{1}{R} \right\} \times \frac{R}{R^2} ds'. \qquad (2.15)$$

An approximation to the integration is achieved by first assuming that the spatial variation of A_{ij} is negligible within each patch, evaluating the integrand at the center of the patch, i.e., r_i, and finally multiplying the resultant by the area of the patch. While performing this operation, it is tacitly assumed that the temporal variation over the patch due to the changing retarded time is negligibly small. Next, it is necessary to numerically interpret the time derivatives appearing in (2.15) in order to complete the evaluation of the integral. One might again assume a pulse-type approximation, i.e., J is constant within a patch, for the temporal variation in a manner similar to the one used for the spatial behavior of A_{ij}. This simple representation can in fact be used if all of the space samples are separated by integral multiples of $c\Delta t$. While such a division may be conveniently employed when the geometry of the scatterer is an extremely simple one, typically one has to deal with the more general situation in which this constraint may be violated. Thus, it often becomes necessary to use an interpolation scheme as outlined below.

One may use, for instance, a second-order interpolation type of representation, viz.,

$$A_{ij}(r_i, t - t_j) = A_{ij}^{(0)}(r_i) + A_{ij}^{(1)}(r_i)(t - t_j) + A_{ij}^{(2)} \cdot (r_i)(t - t_j)^2 \qquad (2.24)$$

or even higher-order interpolation formulas if this is deemed necessary. The second-order interpolation formula above contains three unknown coefficients per patch, viz., $A_{ij}^{(0)}(r_i), A_{ij}^{(1)}(r_i)$, and $A_{ij}^{(2)}(r_i)$. Two of these, $A_{ij}^{(1)}$ and $A_{ij}^{(2)}$, may be expressed in terms of the sampled values of adjacent temporal currents, i.e., $A_{ij}(r_i)$ at $t = t_{j-1}$ and t_{j+1}. This may be done, for instance, by requiring that $A_{ij}(r_i, t - t_j)$ be identically zero at $t = t_{j\pm1}$ as shown in Fig. 2.5. Several alternate interpolation schemes

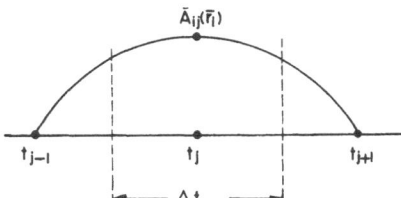

Fig. 2.5. Second order interpolation of expansion coefficients $\bar{A}_{ij}$

are possible, including those based on the use of spline functions [2.19] that have certain interesting and desirable properties.

It should be mentioned that in the interpolation scheme illustrated in Fig. 2.5 there exists the possibility of interpolating to future current values if $t-t_j<2\Delta t$. In this event, the interpolation interval is shifted backwards in time, thereby circumventing the above-mentioned problem. Thus, one may write a general expression

$$A_{ij}(r_i, t-t_j) = \sum_{m=0}^{2} A_{i,j-(v-m)}^{(0)}(r_i) \sum_{n=0}^{2} B_{m,n}(t-t_j)^n \qquad (2.25)$$

where
$$v = 2 \text{ if } (t-t_j)<2\Delta t$$
$$= 1 \text{ if } (t-t_j)>2\Delta t$$

and the coefficients $B_{m,n}$ are determinable from the nature of the interpolation scheme being employed, and hence are considered as known. Note that the index $v=2$ in the summation assures that interpolation does not occur in the future when the time interval $t-t_j<2\Delta t$, but to the two time-samples backwards of t_j. On the other hand, if $t-t_j$ is greater than $2\Delta t$, the interpolation is conventional, i.e., the coefficient A_{ij} is related to the two adjacent coefficients at $t=t_{j+1}$ and $t=t_{j-1}$.

The final step is to substitute the expression (2.25) for $A_{ij}(r_i, t-t_j)$ into (2.23), insert the resultant expression for J_s into (2.15), and perform the integration as indicated in that equation. The required time-derivative $\partial/\partial\tau$ ($=\partial/\partial t$) can be readily obtained by differentiating the polynomial representation in time. The sougth-for equation for the unknowns $A_{ij}^{(0)}$ can now be derived by point-matching in space and time at $r=r_i$ and $t=t_j$, i.e., at the center of time and space intervals. This leads to the equation

$$A_{ij}^{(0)}(r_i,t) = 2\hat{n}_i \times H_{ij}^{\text{inc}} + \frac{1}{2\pi}\hat{n}_i \times \sum_{p=1}^{N_s} \left\{ \sum_{m=0}^{2} A_{p,q-(v-m)}^{(0)} \right. \qquad (2.26)$$

$$\left. \times \sum_{n=0}^{2} B_{m,n} \left[(t'-t_q)^n \frac{1}{R_{iq}^3} + n(t'-t_p)^{n-1} \frac{1}{R_{ip}^2 c} \right] R_{ip} \right\} \Delta s_p$$

where $R_{ip} = r_i - r_p$; $j=1,...N_t$; $i,p=1,...N_s$
$\qquad R_{ip} = |R_{ip}|$; $t'=t-R_{ip}/c\Delta t$; $t=t_0+(j-1)\Delta t$
$\qquad t_0 = $ convenient time origin
$\qquad q = j-r_{ip}$ (not to be confused with charge q)
$\qquad r_{ip} = $ rounded-off values of $R_{ip}/c\Delta t$
$\qquad v = 2$ if $t-t_j<2\Delta t$
$\qquad = 1$ otherwise.

All the quantities in the r.h.s. of (2.26) are known, since, for the latest time sample $t=t_j$, the highest value $q-v-m$ can take is $q=j-r_{ip}$, which is at least one time sample in the past. Recall that the coefficients

$B_{m,n}$ are completely known from the interpolation formula being employed. Of course, the tangential component of the incident field at the i-th patch, i.e., $\hat{n}_i \times H^{inc}_{ij}$ is known for all time. Thus, we may obtain $A^{(0)}_{ij}(r_i, t)$ at the latest time $t = t_j$ for all values of r_i and then proceed to repeat the process for future time steps to derive the complete solution to the problem. One other restriction that should be mentioned here is the choice of the time-stepping interval Δt which should be less than $\Delta r_{min}/c$, with Δr_{min} the minimum value of R_{ip}. It is possible to extend the procedure to the case where the above condition on Δr_{min} is not met; however, this requires the inversion of a matrix equation. We will not go into the details of deriving such a matrix equation here, though a similar situation will be illustrated in connection with the numerical E integral equation for wire structures which is outlined below.

2.3.2 Numerical Solution of the EFIE for Thin Wires

In this Subsection we briefly discuss the numerical solution of the time-domain integral equation (2.22) for thin wires. The equation itself was derived in Subsection 2.2.3 and is repeated here for convenience

$$\hat{s} \cdot E^{inc}(r,t) = \frac{\mu_0}{4\pi} \int \left[\frac{\hat{s} \cdot \hat{s}'}{R} \frac{\partial}{\partial \tau} I(s', \tau) \right.$$

$$\left. + c \frac{\hat{s} \cdot R}{R^2} \frac{\partial}{\partial s'} I(s', \tau) + c^2 \frac{\hat{s} \cdot R}{R^3} \int_{-\infty}^{\tau} \frac{\partial}{\partial s'} I(s' \cdot t') dt' \right] ds',$$

$$r \in C + a. \tag{2.22}$$

The principle on which the method of solution of this equation is based was outlined in Section 2.2. However, it will be worthwhile to examine some of the details involved in implementing the procedure.

Since the kernel of even the one-dimensional E integral equation for wire-type structures has singularities that are of higher order than those present in the MFIE equation investigated in the last section, it becomes necessary to use a smoother type of basis functions for expanding the induced current I as a function of position along the wire. The simple pulse function that was used in connection with MFIE is no longer suitable unless the kernel of (2.22) is first smoothed numerically. We will illustrate the procedure for numerically smoothing the kernel of an integral equation in Subsection 2.3.3 where the use of the E equation for electrically thin surfaces will be described. For the present we discuss the extension of the method described in Subsection

2.3.1 so that it may be employed for kernels that are less smooth than those of the MFIE. The discussion presented below is based in large part on a recent report by VAN BLARICUM [2.20].

The first step subsectionalizes the thin wire by dividing it into a number of segments N_s and defines a set of basis functions for expressing the unknown current I in each of these segments. Similar segmentation is also necessary for time with the resultant expression for $I(s,t)$ which reads

$$I(s,t) = \sum_{i=1}^{N_s} \sum_{j=1}^{N_t} I_{ij}(s-s_i; t-t_j) V(s_i) U(t_j)$$

$$V(s_i) = 1, \quad |s-s_i| \leq \Delta s_i/2$$

$$= 0, \quad \text{otherwise} \tag{2.27}$$

$$U(t_j) = 1, \quad |t-t_j| \leq \Delta t/2$$

$$= 0, \quad \text{otherwise}$$

where, similar to the notation in (2.23), Δs_i and Δt_j $(=\Delta t)$ are the lengths of the spatial segment i and time interval j; the latter is assumed to be constant and independent of j. It should be pointed out that the curved wire is approximated by straight segments (see Fig. 2.6)

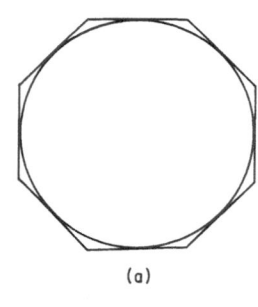

(a)

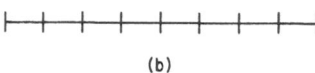

(b)

Fig. 2.6a and b. Example of a curved wire viz., a loop, modeled by straight sections

just as a general surface structure is approximated by planar patches. The second step uses an interpolation scheme to express the current in one space and time segment in terms of the current values in the neighboring segments. The reader may recall that this procedure was employed in the last section for the temporal variable only. Due to the higher-order singularity in the kernel of E equation (2.22), it becomes necessary to employ similar interpolation in the spatial variable as well. As before, we may use a second-order polynomial representation with two of the three coefficients determined by the interpolation scheme. For the sake of illustrating the procedure, we will agin choose the interpolation to be Lagrangian, though other schemes are also possible.

For a fixed t we may write

$$
\begin{aligned}
I_{ij}(s-s_i; t-t_j) = {} & \frac{(s_i-s)(s_{i+1}-s)}{(s_i-s_{i-1})(s_{i+1}-s_{i-1})} I_{i-1}(t-t_j) \\
& + \frac{(s_{i-1}-s)(s_{i+1}-s)}{(s_{i-1}-s_i)(s_{i+1}-s_i)} I_i(t-t_j) \\
& + \frac{(s_{i-1}-s)}{(s_{i-1}-s_{i+1})} \frac{(s_i-s)}{(s_i-s_{i+1})} I_{i+1}(t-t_j).
\end{aligned}
\tag{2.28}
$$

The $I_i(t-t_j)$ terms may also be expanded in a similar manner

$$
\begin{aligned}
I_i(t-t_j) = {} & \frac{(t_j-t)(t_{j+1}-t)}{(t_j-t_{j-1})(t_{j+1}-t_{j-1})} I_{i,j-1} \\
& + \frac{(t_{j-1}-t)(t_{j+1}-t)}{(t_{j-1}-t_j)(t_{j+1}-t_j)} I_{i,j} \\
& + \frac{(t_{j-1}-t)(t_j-t)}{(t_{j-1}-t_{j+1})(t_j-t_{j+1})} I_{i,j+1},
\end{aligned}
\tag{2.29}
$$

where the $I_{i,j}$ are the terms representing the magnitude of the current at the center of the i-th space segment and the j-th time segment.

A combined form of expansion if I_{ij} in space and time may now be constructed. It may be written as

$$
\begin{aligned}
I_{ij}(s-s_i; t-t_j) &= \sum_{l=-1}^{1} \sum_{m=v}^{v+2} B_{ij}^{(l,m)} I_{i+l, j+m} \\
B_{ij}^{(l,m)} &= \prod_{p=-1}^{p=1\,(l)} \prod_{q=v}^{q=v+2\,(m)} \frac{(s-s_{i+p})(t-t_{j+q})}{(s_{i+l}-s_{i+p})(t_{j+m}-t_{j+q})},
\end{aligned}
\tag{2.30}
$$

where the superscripts (l) and (m) on the product symbol indicate that $p=l$ and $q=m$ are to be omitted from the products. As in the last example we assume that $|s-s_i| < \Delta s_i/2$ and $|t-t_j| \le \Delta t/2$. Also, in order to avoid extrapolation into the future, we set $v=-2$ if $\Delta R=$

$R/[c(t_j-t_{j-1})] < 0.5$ for the j-th time step, for which the unknown current is being computed. If $\Delta R > 0.5$, v is set equal to -1. In the latter case the temporal interpolation of the current is carried out using its values at the two adjacent time steps, one forward and one backward.

The third step expresses the integral w.r.t. time, appearing in (2.22), in a form convenient for numerical handling. This is done by writing

$$-\int_{-\infty}^{\tau} \frac{\partial I}{\partial s}(s,t')dt' = q(s',\tau)$$

$$= \sum_{i=1}^{N_s} \sum_{j=1}^{N_t} q_{ij}(s'-s_i; \tau-\tau_j) V(s_i) U(\tau_j) \qquad (2.31)$$

where

$$q_{ij}(s'-s; \tau-\tau_j) = \sum_{l=-1}^{l=1} \sum_{m=v}^{m=v+2} B_{ij}^{(1,m)} q_{i+l,j+m}$$

and the coefficients in the above expansion are given by

$$q_{i,j} = -\sum_{k=1}^{k=j} \frac{\partial}{\partial s''} \int_{\Delta t} I_{ik}(s-s_i; t-t_j) dt'' \qquad (2.32)$$

with $s''=s'-s_i$, and $t''=\tau-\tau_j$.

The fourth and last step introduces the expansions (2.30) and (2.31) for the current and the charge into the integral equation (2.22) and applies point matching to generate the desired matrix equation. Since the wire structure has been modeled by N_s straight-wire segments, the integrations along the wire are also approximated accordingly. In carrying out these integrations, we may note some important characteristics of the form of representation we have chosen for the unknown current distribution I. Referring to (2.30), we note, for instance, that the evaluation of the self-patch integral at $s=s_i$ at the *latest* time segment, e.g., T_M, will involve nine coefficients, viz., $I_{i,M}$, $I_{i-1,M}$, $I_{i+1,M}$, $I_{i,M-1}$, $I_{i-1,M-1}$, $I_{i+1,M-1}$ and $I_{i,M-2}$, $I_{i-1,M-2}$, $I_{i+1,M-2}$, because the interpolation in space couples the current I_i at space sample s_i to its two adjacent neighbors, i.e., I_{i+1} and I_{i-1}. The same would normally be true for interpolation in time. However, since it is necessary to avoid interpolation into the furure, I_M is coupled to its counterparts at the two backward time steps, viz., t_{M-1} and t_{M-2}. This coupling is done by setting $v=-2$ in (2.30), which is consistent with the fact that ΔR is less than <0.5 in the self-patch. Next, we note that for all of the other integrals, i.e., the non-self-patch ones, the delayed time $\tau=t_M-R_i/c$ will be greater than or equal to one time sample, and hence the temporal index j for the current coefficients $I_{i,j}$ appearing in the result of these integrals will be $M-1$ or less. If all of the $I_{i,j}$ have been determined previously for $j \leq M-1$, then the unknown coefficients $I_{i,M}$ appear only in the evaluation of the self-patch integrals.

It may be instructive at this stage to write the numerically approximated version of the integral equation (2.22) after it has been point-matched at $s = s_u$, and $t = t_v$. We get

$$\hat{s}_u \cdot E^{\text{inc}}(s_u, t_v) = \frac{\mu_0}{4\pi} \sum_{i=1}^{N_s} \hat{s}_u \cdot \int_{\Delta c_i} \left[\hat{s}_i \frac{\partial I_{ij}(s' - s_i; t_v - t_j - R_{iu/c})}{\partial t_v} \cdot \frac{1}{R_{iu}} \right.$$

$$+ c \frac{R_{iu}}{R_{iu}^2} \frac{\partial I_{ij}(s' - s_i; t_v - t_j - R_{iu/c})}{\partial s'}$$

$$\left. - c^2 \frac{R_{iu}}{R_{iu}^3} q_{ij}(s' - s_i; t_v - t_j - R_{iu/c}) \right] ds'. \qquad (2.33)$$

It should be pointed out that q_{ij}'s appearing in (2.33) are not additional unknowns as they have already been related to I_{ij}'s via (2.32).

For the self-patch integral we have $r_u = r_i$ and $R_{iu}/c < \Delta t/2$, i.e., less than half the time step and, hence, we round R_{iu}/c off to zero. At the observation time $t_v = t_M$, i.e., at the latest time sample, the maximum value t_j can take is also t_M, yielding $t_v - t_j - R_{iu}/c \approx 0$. For all other r_u or r_i such that $r_u \neq r_i$, $t_v - t_j - R_{iu}/c$ is at least Δt or larger and t_j is $t_v - \Delta t$ or less. Since the values of I_{ij} at past times are known from previous computation, the only unknowns involved in the computation of the currents at $t = t_M$ are $I_{i+1,M}$, $I_{i-1,M}$ and $I_{i,M}$.

The matrix version of (2.22) may be obtained by substituting the expressions for I_{ij} and q_{ij}, given earlier, into the numerically-approximated form of the integral equation. The resulting equation has a tridiagonal form and its inversion can be obtained rather efficiently using standard numerical techniques for inverting such matrices. The algebraic manipulations required to derive the matrix equation are rather straightforward, but the form of the resultant equation is rather involved. The reader interested in the details of this equation may choose to consult [2.20 or 2.14].

The final form of the matrix equation may be written symbolically as

$$[D]\{I_M\} = \{E^{\text{inc}}|_{t_v = t_M}\} + \{F\}$$

$$\{F\} = \text{column vector known from previous computation}$$

$$(2.34)$$

the elements of $\{I_M\}$, i.e., $I_{i,M}$, are the unknowns which we are attempting to determine. The first term in the r.h.s. of (2.34) is a column vector whose elements are E^{inc} evaluated at the latest time $t_v = t_M$ and observation segment $s = s_u$. The second term in the r.h.s. contains only known quantities that have been obtained during previous computations. Thus, one has only to invert the tridiagonal matrix D, all of whose elements are known in terms of interpolation coefficients, etc., to compute the elements $I_{i,M}$ for all i at $t_v = t_M$. It should be mentioned that the matrix inversion is required to be carried out only once and does not have to be redone for each future time-stepping, when the computations are performed at $t = t_{M+1}$, t_{M+2}, etc., because the elements of the matrix D are independent of the temporal variation t and depend only on the geometry of the wire.

Before concluding this section, it will be useful to point out that in this chapter we have presented the details of only one of the several possible approaches for solving the thin-wire integral equation arising in time-domain scattering problems. Another procedure, which is based on the finite-difference approach as outlined in Subsection 2.3.3 rather than the interpolation procedure described in Subsection 2.3.2, is discussed Chapter 4 and has also been investigated by AUCKENTHALER and BENNETT [2.9]. As mentioned earlier, an interpolated form of representation for the current is not necessary after the kernel has been smoothed via finite-difference representation. It has been found [2.21], however, that extreme care must be exercised in using the finite difference form of approximation when the wire geometry contains junctions or bends. Also, the finite difference formula has been known to produce difficulties when applied to solid-surface geometries that contain bends or edges where the surface current distribution may possess singularities. However, these difficulties can be successfully circumvented by a proper application [2.11, 21] of the finite-difference procedure.

2.3.3 Finite Difference Approach for EFIE

In this Subsection we describe an alternate approach to numerical solution of the EFIE, one that is based on finite difference interpretation of the differential operators appearing in the time-domain integro-differential equation. As mentioned earlier, the finite-difference interpretation has the primary effect of smoothing the kernel of the equation of interest, viz., (2.13). This, in turn, reduces the order of the singularity of the kernel and allows the use of simple basis functions, e.g., pulse functions for the expansion of the unknown surface current density J_s. Recall that this was also possible with the MFIE whose kernel is less singular than the EFIE. Another feature of the finite difference approach is that it provides a convenient means for deriving a time-stepping algorithm for solving EFIE. This fact will be apparent from the discussion to be presented below. Incidentally, as will shortly be seen, the spatial and temporal derivatives may both be interpreted in terms of finite-difference operators.

The discussion will follow somewhat along the lines described by BENNETT et al. [2.11]; it will be found convenient to start with a slightly different form of the EFIE, one written in terms of the vector potential A.

The scattered electric field, produced by the induced current, may be written as

$$E = -\frac{\partial A}{\partial t} - \nabla \phi. \tag{2.35}$$

Taking the tangential component of this equation, and equating $\hat{n} \times E$ to $-E^{\text{inc}}$, we have after expressing ϕ in terms of A via the gauge condition,

$$\hat{n} \times E^{\text{inc}} = \hat{n} \times \frac{\partial A}{\partial t} - \frac{1}{c^2} \int^t \hat{n} \times \nabla \nabla \cdot A \, dt', \quad r \in S. \tag{2.36}$$

The integral w.r.t. time in (2.36) may be eliminated by differentiating the entire equation w.r.t. t, yielding

$$\hat{n} \times \frac{\partial E^{\text{inc}}}{\partial t} = \hat{n} \times \frac{\partial^2 A}{\partial t^2} - \frac{1}{c^2} \hat{n} \times \nabla \nabla \cdot A, \quad r \in S. \tag{2.37}$$

We use the expression (2.9b) for the vector potential A in terms of the induced surface current J_s to obtain

$$A(r,t) = \frac{\mu_0}{4\pi} \int_S \frac{J_s(r', t - R/c)}{R} \, ds' \tag{2.38}$$

by restricting J to be a surface distribution. From (2.37) and (2.38) we finally derive (see also Subsect. 4.6.2).

$$\frac{\mu_0}{4\pi} \hat{n} \times \frac{\partial^2}{\partial t^2} \int_S \frac{J_s(r', \tau)}{R} \, ds' - \frac{\hat{n}}{4\pi\varepsilon_0} \times \nabla \nabla \cdot \int_S \frac{J_s(r', \tau)}{R} \, ds' = \hat{n} \times \frac{\partial E^{\text{inc}}}{\partial t},$$

$$r \in S; \quad \tau = t - R/c = \text{retarded time}. \tag{2.39}$$

Although analytically (2.39) is equivalent to the EFIE (2.13), there are subtle differences in the manner in which the derivative operators appear in the two equations. This, in turn, has a profound influence on their numerical interpretation as explained below. Notice first of all that both the temporal and spatial derivatives appear *outside* the integrals, i.e., in r and t coordinates, in contrast to the inside derivatives in (2.13). This fact will be significant when these derivatives are interpreted numerically. A second point to note is the conspicuous absence of the P.V. sign on the integrals. The reason is that, in contrast to (2.13), the kernels of the integrals in (2.39) are much less singular, and are in fact integrable. Hence special interpretation in terms of the P.V. operator is no longer necessary here. The third point to note is the absence of the charge density ϱ inside the integral, because by taking the time-derivative of the entire equation we have been able to write the entire equation in terms of J_s.

Our next objective is to rewrite (2.39) in a form suitable for iterative solution by time stepping. To illustrate the fundamental concept on which the procedure is based let us consider a planar surface for which the spatial derivatives can be written in simple forms.

The quantity $\nabla\nabla \cdot J_s$ may be written explicitly as

$$\hat{n} \times \nabla\nabla \cdot J_s = \hat{y}\left(\frac{\partial^2 J_{sx}}{\partial x^2} + \frac{\partial^2 J_{sy}}{\partial x \partial y}\right) = \hat{x}\left(\frac{\partial^2 J_{sx}}{\partial x \partial y} + \frac{\partial^2 J_{sy}}{\partial y^2}\right) \qquad (2.40)$$

where we have assumed that the normal to the surface is along the z-direction. For a given point x_m, y_m on the surface, one can write the following finite difference interpretations, operating on a function $F(x,y,t)$

$$\left.\frac{\partial^2 F(x,y,t)}{\partial x^2}\right|_{\substack{x=x_m \\ y=y_n \\ t=t_j}} = \frac{F(x_{m+1}, y_n; t_j) - 2F(x_m, y_n; t_j) + F(x_{m-1}, y_n; t_j)}{(\Delta x)^2} \qquad (2.41)$$

which is the standard central difference form for the derivative operator $\partial^2/\partial x^2$ with $\Delta x = x_p - x_{p-1}$, the grid spacing in the x-direction. Likewise for $\partial^2/\partial y^2$ we may write

$$\left.\frac{\partial^2 F(x,y,t)}{\partial y^2}\right|_{\substack{x=x_m \\ y=y_n \\ t=t_j}} = \frac{F(x_m, y_{n+1}; t_j) - 2F(x_m, y_n; t_j) + F(x_m, y_{n-1}; t_j)}{(\Delta y)^2}. \qquad (2.42)$$

Δy is the grid spacing in the y direction, and is typically set equal to Δx. For the mixed derivative we have

$$
\left.\frac{\partial^2 F}{\partial x \partial y}\right|_{\substack{x=x_m \\ y=y_n \\ t=t_j}} = \frac{1}{4\Delta x \Delta y} \left[F(x_{m+1}, y_{n+1}, t_j) + F(x_{m-1}, y_{n-1}, t_j) \right.
$$
$$
\left. - F(x_{m+1}, y_{n-1}, t_j) - F(x_{m-1}, y_{n+1}, t_j) \right]. \tag{2.43}
$$

Finally, for the temporal derivative $\partial^2/\partial t^2$, we write

$$
\left.\frac{\partial^2 F}{\partial t^2}\right|_{\substack{x=x_m \\ y=y_n \\ t=t_j}} = \frac{F(x_m, y_n, t_{j+1}) - 2F(x_m, y_n, t_j) + F(x_m, y_n, t_{j-1})}{(\Delta t)^2} \tag{2.44}
$$

where Δt is the width of the time-sampling period. It is evident that the above expressions, which provide numerical approximations of the derivatives, tend to be exact only in the limit Δx, Δy and Δt all approaching 0.

The key point to note here is that the temporal derivative $\partial^2/\partial t^2$ contains the time sample at $(t_j + \Delta t)$, i.e., $F(x_m, y_n, t_{j+1})$. Using this fact we can express this quantity in terms of the time samples at previous times and thereby derive a formula that is suitable for time-stepping. To this end we rewrite (2.39) as

$$
\hat{n} \times F|_{t=t_M} = (\Delta t)^2 \left[\frac{4\pi}{\mu_0} \frac{\partial}{\partial t} (\hat{n} \times E^{\text{inc}}) + \frac{\hat{n}}{4\pi\varepsilon_0} \times \nabla\nabla \cdot F \right]_{t=t_{M-1}} \tag{2.45}
$$
$$
+ 2\hat{n} \times F|_{t=t_{M-1}} - \hat{n} \times F|_{t=t_{M-2}}
$$

where

$$
F(r, t) = \int_S \frac{J_s(r', \tau)}{R} ds' = \frac{4\pi}{\mu_0} A. \tag{2.46}
$$

Note that the operator $\nabla\nabla\cdot$ in (2.45) may be replaced by its finite difference interpretations given in (2.41) through (2.43).

The important thing to note in (2.45) is that $(\hat{n} \times F)$ at the latest time sample, say t_M, has been expressed in terms of quantities evaluated at two previous samples at t_{M-1} and t_{M-2}, and in terms of the time derivative of the incident field which is known for all time. Thus (2.45) allows us to compute the values of F at future times from the present and past values of F, assuming that the latter are known. However, by itself, this equation is not as useful as might appear at first sight, because our primary goal is to compute the induced current density J_s which is not obtainable from F in an obvious manner from (2.46) that relates these two quantities. We can, nevertheless, make advantageous use of this relationship to derive a time-stepping procedure directly for the induced current J_s rather than F. The clue to deriving this formula lies in breaking up the integral representation in (2.46) into the following two parts

$$F(r,t) = \int_\Sigma \frac{J_s(r',\tau)}{R} ds' + \int_{S-\Sigma} \frac{J_s(r',\tau)}{R} ds' \tag{2.47}$$

where Σ is a small patch, $\delta \times \delta$ in size, enclosing the point $r = r'$, and δ is the sample size in the x and y directions. We recognize the first integral as the self-patch integral, and if the patch size is chosen such that $\delta/c = \Delta t$, then $t - \Delta t < \tau < t$ within Σ. Thus, τ may be rounded off to t within the self patch. Next, if we make a pulse approximation for J_s, which is permissible since the kernel has only a weak singularity, we may approximate the self-patch integral to give

$$\int_\Sigma \frac{J_s(r,\tau)}{R} ds' \approx \alpha J_s(r,t) \tag{2.48}$$

where

$$\alpha = 4\delta \ln(1 + \sqrt{2}). \tag{2.49}$$

Next, we study the second integral in the r.h.s. of (2.47). We note that the time appears in this integral through τ and, in contrast to the self-patch integral, the temporal argument of the second integral is *always* retarded. Thus, we have the result

$$F(r,t) = \alpha J_s(r,t) + \text{terms containing } J_s \text{ evaluated at retarded times.} \tag{2.50}$$

Using (2.50) in (2.45) we may write the latter as

$$\alpha J_s(r,t)|_{t=t_M} = (\Delta t)^2 \left(\frac{4\pi}{\mu_0} \frac{\partial}{\partial t} \hat{n} \times E^{\text{inc}} \right) \tag{2.51}$$

$$+ \text{terms containing } J_s \text{ at } t = t_{M-1} \text{ or less}.$$

Equation (2.51) is the result we were seeking. Initially, at the first time step $t_1 (= \Delta t)$ we have from (2.51)

$$\alpha \, \boldsymbol{J}_{\mathrm{s}}(\boldsymbol{r},t)\big|_{t=t_1} = (\varDelta t)^2 \left(\frac{4\pi}{\mu_0} \, \frac{\partial}{\partial t} \, \hat{\boldsymbol{n}} \times \boldsymbol{E}^{\mathrm{inc}} \right)_{t=t_1} \tag{2.52}$$

since all of the currents are identically zero everywhere at $t=0$. It is now a straightforward matter to step on time and iteratively compute $\boldsymbol{J}_{\mathrm{s}}$ at $t=t_2, t_3, \ldots$, etc., up to any desired time. The terms in the r.h.s. of (2.51) containing $\boldsymbol{J}_{\mathrm{s}}$ at retarded time will, of course, have non-zero contribution at future time, but they may be evaluated using numerical integration in conjunction with finite difference approximation upon substitution of the known values of $\boldsymbol{J}_{\mathrm{s}}$ already computed for previous times.

Before closing this subsection it may be worthwhile to reiterate that special consideration for finite differencing with respect to the spatial coordinates may be necessary [2.11] at sharp edges where one of the current components may have a singular behavior. It should also be mentioned that the $\nabla\nabla\cdot$ operator may be approximated in curvilinear coordinates as well, and that the finite-difference method itself is not restricted to planar geometries only.

2.3.4 Numerical Considerations

In this subsection we consider certain questions that arise in the numerical solution of various time-domain integral equations that have just been discussed. Specifically, we consider the aspects of accuracy, beginning and terminating times for the solution, the influence of symmetry, the choice of excitation waveform, and other items of this nature.

Clearly, the accuracy of the numerical solution will depend on how well the integral equation describing a problem has been approximated by numerical means. The sampling rates in the spatial and temporal domains are key factors in determining the accuracy of the numerical approximation. The shape of the exciting pulse and the ratio of body size to the effective width of the pulse in space also have important influence on the accuracy of the solution, as will be evident from the discussion given below.

There are no unique guidelines for spatial sampling of a given geometry but roughly speaking for a cw illumination of wavelength λ, $\varDelta R \sim \lambda/N$, where N is order of 4 to 8 for the surface integral equation and 6 to 20 for the wire integral equation. The minimum rate of sampling can be considerably larger for curved wires, e.g., a spiral, than for a straight wire, an L-shaped wire, or even a moderately curved wire. For the surface structure the sampling rate depends upon the type of integral equation used, and whether corners or edges are present on the structure that might produce rapid variations in the induced current. Further discussions on spatial sampling requirements may be found in [2.7, 14]; however, for the purposes of the following discussion we will assume that the guideline for N mentioned above is adequately satisfactory.

We turn next to the question of temporal sampling. In order for the time-stepping algorithms described earlier in this section to work without the need of matrix inversion, it is necessary that we choose $\Delta t \sim \Delta R/c$. If Δt is chosen to be greater than $\Delta R/c$, then there will be coupling between the various adjacent patches and the resultant equation for the latest time sample for the induced current will have mutual coupling terms. This will destroy the diagonal nature that was obtained for the MFIE discussed in Subsection 2.3.1, and for the finite-difference approach described in Subsection 2.3.3. If $\Delta t = 2R/c$, for instance, the resultant matrix will exhibit coupling effects to each of the patches in the immediate neighborhood of the self-patch. In common with the interpolation approach presented in Subsection 2.3.2, a matrix inversion will then be necessary to construct the solution, although the matrix will be sparse and the inversion has to be carried out only once. Nevertheless, the procedure will be more involved than in the case when Δt is set equal to $\Delta R/c$.

There is one other consideration that points toward a higher sampling interval in time than $\Delta R/c$. If we use the sampling theorem as our guide, we find that the minimum temporal sampling rate for a band-limited signal is $\Delta t = 1/(2f_{max})$, where f_{max} is the effective cutoff frequency of the input signal. If we assume that the maximum effective frequency content of the input pulse is f_{max}, then the same upper frequency bound will usually hold also for the induced current. Thus, on the basis of the sampling theorem argument we can set the time sampling interval to be $1/(2f_{max})$, where f_{max} is as defined above. However, according to the spatial sampling guide, a monochromatic source of frequency f_{max} would require $\Delta R = \lambda/N = c/(f_{max}N)$ and the corresponding Δt will be $1/(f_{max}N)$. Using the guidelines for N given above we get

$$\Delta t \sim \frac{1}{6 f_{max}} \quad \text{to} \quad \frac{1}{20 f_{max}} \quad \text{for wires}$$

$$\text{and} \quad \Delta t \sim \frac{1}{4 f_{max}} \quad \text{to} \quad \frac{1}{8 f_{max}} \quad \text{for surfaces}.$$

It is obvious that the above sampling widths are several times smaller than those required by the sampling theorem guidelines. We are therefore faced with a tradeoff situation, namely that we have the possibility of enhancing the temporal sampling efficiency by increasing t, but only at the cost of having to carry out a matrix inversion.

Next we turn to the question of how the shape of the exciting pulse might affect the numerical computation. In principle, a single time-domain computation with an arbitrary exciting pulse can provide the response characteristics of the target for all frequencies. However, in practice the high frequency information is typically corrupted by noise introduced by the numerical processing. To minimize this effect, it is desirable to work with an input pulse that has the widest possible bandwidth, one approaching the delta function. Numerically, this is accomplished by working with a Gaussian pulse $g(t)$ given by

$$g(t) = e^{-t^2 A^2}$$

whose frequency spectrum $G(f)$ is given by

$$G(f) = e^{-\pi^2 f^2/A^2}.$$

At $f = 0.5\,A$, $G(f) \approx 0.1$. If we define the effective bandwidth on this basis, we have $f_{max} = 0.5\,A$. According to the guidelines we have developed, we would choose $\Delta t \sim 1/(6f_{max}) = 1/3\,A$. Since the input pulse is down to 0.1 of its value at $t \approx 1.5/A$, the above choice of t implies that the pulse will be sampled approximately 5 times. It is necessary to ascertain that this sampling rate is adequate to accurately represent the input pulse. Of course, decreasing the spatial sampling has the effect of increasing the number of samples in the time domain because of the reduced sampling interval.

There is yet another consideration that must be taken into account when choosing the pulse width parameter A. This is the width of the pulse relative to the body size. This point will now be discussed in more detail.

The width of the pulse determines its effective maximum frequency f_{max}. According to the discussion in Chapter 3, the scattering or radiating characteristics of a structure can be described in terms of its resonant singularities or poles in the complex frequency plane (s-plane). The number of these poles that will effectively contribute to the induced current, and hence to the radiated or scattered field, is in turn determined by f_{max} of the exciting pulse. The locations of the poles in the s-plane scale directly with the characteristic length of the body. Thus, increasing f_{max}, which is inversely proportional to the pulse width, allows one to obtain more accurate information on the nature of the scatterer or the antenna. On the other hand, increasing f_{max}, which requires the use of a smaller width for the spatial and temporal samples, consequently increases the computer storage and execution times. Thus, once again, a tradeoff is necessary between increased computation cost and accurate high-frequency information.

Finally, let us briefly consider the criteria for starting and stopping the solution of the time-domain integral equation solution. Typically one starts the solution when the source strength is 10^{-3} to 10^{-2} of its maximum value. The currents and charges both are set equal to zero prior to this time. The stopping criterion is not defined as easily, however. If the structure under consideration has resonances, the induced current waveform may not decay for a long period of time. This will happen when one of its characteristic poles is close to the imaginary axis in the s-plane. In this event, it may be necessary to continue the computation for many time steps, unless a procedure is used to determine the complex frequencies or poles from the early part of the calculation; this information is utilized to obtain the late time information via extrapolation. One such technique, which has recently been introduced by MITTRA and VAN BLARICUM, [2.22] will be described in Section 2.6 in some detail.

Another technique that can be exploited for enhancing the efficiency of computation is the use of symmetry when the structure possesses mirror or rotational symmetries. Effective use of symmetry can reduce the computation time and storage quite substantially. More details on this point may be found in [2.14].

2.4 Representative Results

A rather exhaustive body of literature [2.7, 14, 20, 22] exists for reporting numerical results derived via the use of time-domain integral equations. In this section we present a few selected representative results to illustrate the application of the various techniques described in this chapter.

We begin with some results for simple wire structures, e.g., the straight wire and the single loop. First, let us assume that the straight wire is excited in an antenna mode by an incident pulse $\mathscr{E}(t)$

$$\mathscr{E}(t) = \frac{V_i}{\Delta s_i} \exp\left[-A^2(t - t_{max})^2\right] \tag{2.53}$$

where A = spread parameter of Gaussian pulse
t_{max} = time at which it reaches its maximum value
V_i = maximum applied voltage
Δs_i = length of the segment to which the incident field is applied.

VAN BLARICUM [2.20] and others have computed the induced current as a function of time using the one-dimensional version of the time-domain integral equation with a current interpolation scheme described in Subsection 2.3.2. These induced currents have been used to find the time-dependent radiated fields. The far scattered fields are obtained by letting $R \to \infty$ in the representation for the E and H fields given in (2.11a) and (2.11b). They are given by (see footnote 2 on p. 84)

$$E(r,t) = -\frac{1}{4\pi r c} \int_s \left[\left| \sqrt{\frac{\mu_0}{\varepsilon_0}} \frac{\partial}{\partial \tau} J_s(r',\tau) - \frac{r}{\varepsilon_0 r} \frac{\partial}{\partial \tau} \sigma(r',\tau) \right] ds' \quad (2.54\,a)$$

and

$$H(r,t) = \frac{1}{4\pi r c} \int_s \frac{\partial}{\partial \tau} J_s(r',\tau) \times \frac{r}{r} ds'. \quad (2.54\,b)$$

The input impedance of the antenna may also be computed as a function of frequency from the time-domain solution. This computation is done by Fourier transforming the current waveform at the source segment and dividing the result by the transform of the input wave form.

For numerical computation a 1-meter wire has been considered with the two center sections excited by the input pulse given in (2.53). The wire was divided into 48 segments with $\Delta s = 0.02083$ m. The radius of the wire was 0.00764 m. The magnitude of the Gaussian pulse, which was applied at the center of each of the two segments, was 0.5 V. The spread parameter A of the Gaussian pulse was chosen to be $3.52 \times 10^{-9}\,\text{s}^{-1}$, and t_{max} was selected so that maximum voltage was obtained at the twentieth-time step. Each time step Δt was uniformly chosen to satisfy $\Delta t = \Delta s/c = 6.945 \times 10^{-12}$ s. The induced current was calculated for a total of 480 time steps and extrapolated to 1024 time steps which were then Fourier transformed using the FFT algorithm. Considerable saving in computer time can be achieved if such an extrapolating procedure is employed to extend the initial temporal range over which the current is computed. As mentioned earlier, a technique for carrying out this type of extrapolation will be described in the next section. Figure 2.7 shows the values of the induced current as a function of time at the center of one of the source segments. One may conjecture from the nature of the waveform that it is a superposition of a number of damped sinusoids. That this is indeed the case will be demonstrated in Section 2.6.

By Fourier transforming the above waveform, one may compute the real and imaginary parts of the input admittance as a function

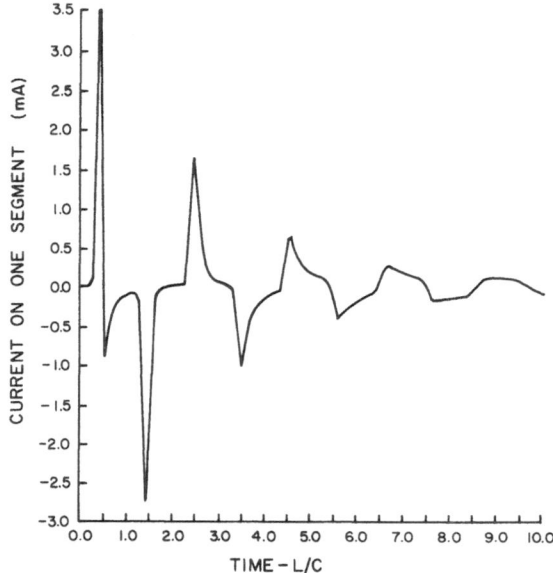

Fig. 2.7. Induced current at the center of one of the two source segments of the linear dipole

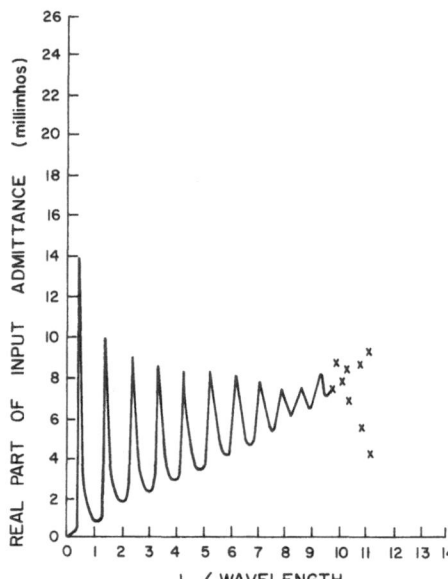

Fig. 2.8. The real part of the input admittance for the linear dipole

of L/λ where the results are given in Figs. 2.8 and 2.9. These plots show that numerical instabilities limit the accuracy at high frequencies above $L/\lambda \sim 10$ (see also Figs. 4.15 and 4.16).

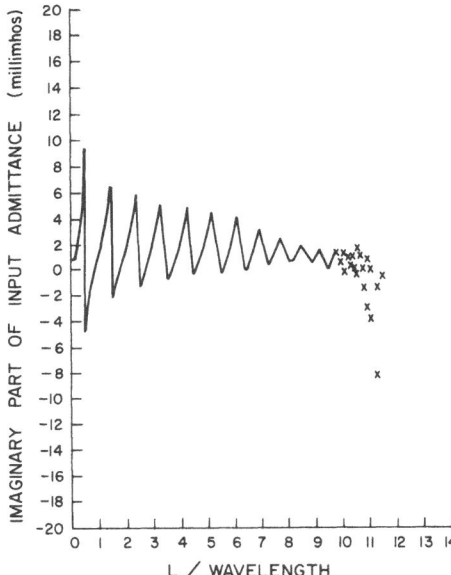

Fig. 2.9. The imaginary part of the input admittance for the linear dipole

The time-domain program can also be used to solve the problem for a pulsed plane wave incident on the wire with only a minor modification. For broadside incidence the incident field is simultaneously applied to all of the segments rather than just to the center ones as in the antenna case. The back-scattered field from a meter-length wire illuminated by the same Gaussian pulse (2.53) as in the antenna case is shown in Fig. 2.10. The presence of damped sinusoidal waves is again evident in the scattered-field waveform.

The wire program has also been applied to a curved-wire structure, viz., the ring resonator. The structure is a single ring-shaped wire with a circumference of 1 m and a wire radius of 0.00674 m. The ring was approximated by 48 straight segments whose centers fell on the perimeter of the circle. The time step Δt was again chosen to satisfy $\Delta t = \Delta s/c$.

The scattering characteristics of the ring resonator for the case of axial incidence were computed for a Gaussian pulse with a spread parameter $A = 3.25 \times 10^9 \, \mathrm{s}^{-1}$. The scattered field was initially calculated

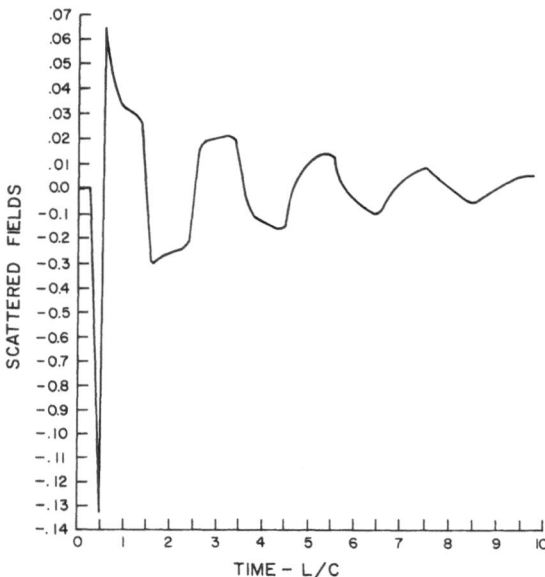

Fig. 2.10. The time-dependent back-scattered fields of the linear dipole for broadside incidence

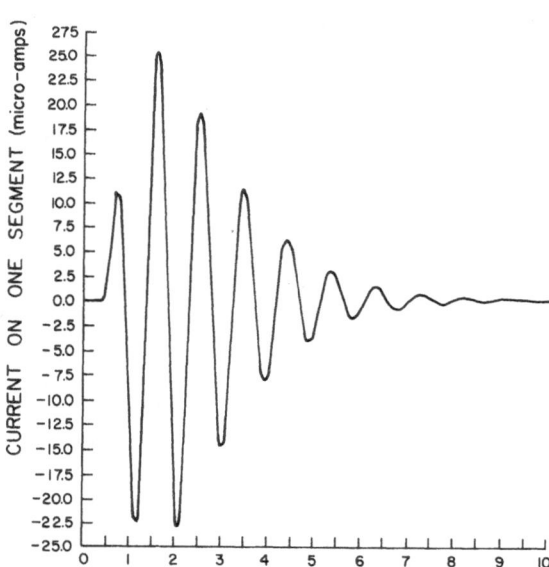

Fig. 2.11. Induced current on one segment of the circular ring scatterer

for 480 time steps and then extrapolated to 1024 time steps. Figure 2.11 shows the current waveform on one of the 48 segments as a function of time and the back-scattered field from the ring is shown in Fig. 2.12. The resonance effect of the ring is evident in both the waveforms. In fact some of the complex exponents of these waveforms lie very close to the imaginary axis in the complex s-plane.

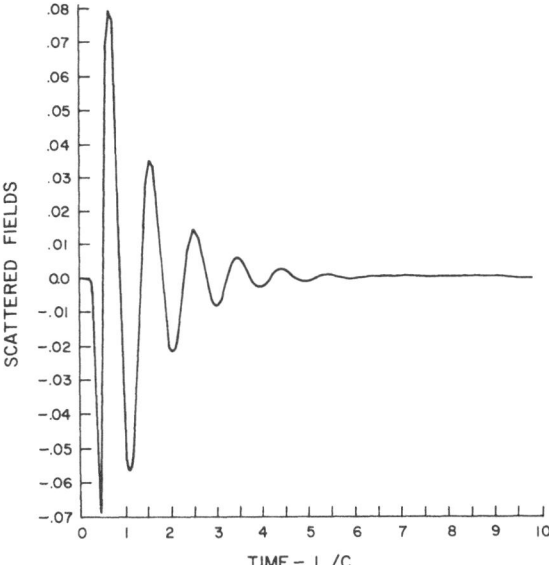

Fig. 2.12. The time-dependent back-scattered fields of the circular ring scatterer for axial incidence

As an example of a two-dimensional structure we present some results for a corner reflector, which has been investigated by BENNETT and WEEKS [2.6] using the appropriate equation for cylindrical structures discussed in Subsection 2.2.2. The geometry of the corner reflector is shown in Fig. 2.13. The subdivisions of the surface used for the numerical solution of the integral equation also are shown there. The impulse response of the corner reflector has been computed for both backside and frontside incidence. These are shown in Figs. 2.14 and 2.15 for the TE case. Note that the field scattered in the shadow region is quite similar in both cases; however, there is substantial difference in the waveforms scattered in the forward direction even at angles that are quite oblique. In general, the waveforms in both the forward

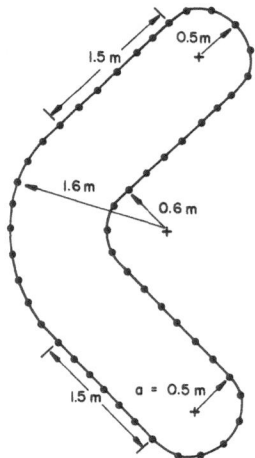

Fig. 2.13. Geometry of the corner reflector

and backward directions are rather complicated for the corner reflector showing the effects of multiple reflection. Although not shown here, there is also a substantial difference in the impulse response for the TE and TM case, see [2.6].

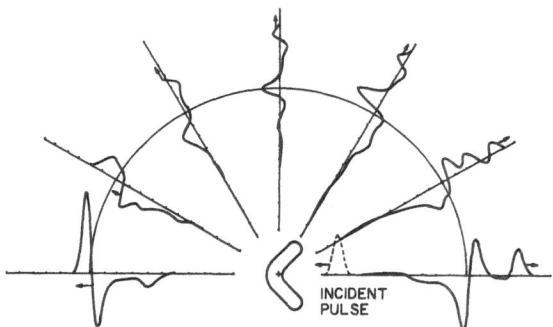

Fig. 2.14. Approximate TE impulse response of a corner reflector with frontside incidence

Finally, we turn to three-dimensional structures. We begin with solid-surface scatterers, such as the sphere. For this example the sphere diameter 2a, which was 1 m, was subjected to a Gaussian field of approximately the sphere's diameter in width and incident along the positive z-axis of a coordinate system centered in the sphere, i.e., $H^{inc} \propto \exp[-A^2(t-z/c)^2]$. The surface sample points had nearly a uniform sepa-

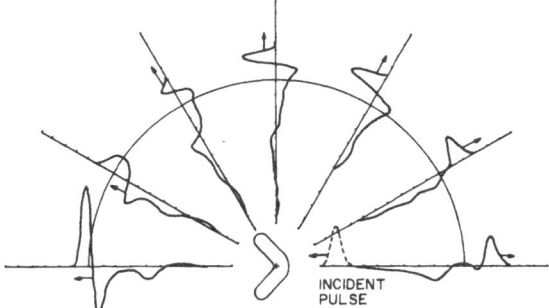

Fig. 2.15. Approximate TE impulse response of a corner reflector with backside incidence

ration of approximately 0.25 m. This separation was accomplished by dividing the sphere into 12 bands of 15 degrees width in elevation angle, and spacing the azimuth samples within each band in multiples of 4 to obtain quadrant symmetry. A sequence of 4, 8, 12, 16, 20, 24, 20, 16, 12, 8, 4 patches per band was obtained for a total of 168. A constant time increment Δt of 0.2 light meter was chosen for the entire calculation. Figure 2.16 shows a comparison of the backscatter results obtained numerically by the integral equation approach using the MFIE and those derived by Fourier transforming the Mie series[3] solution for

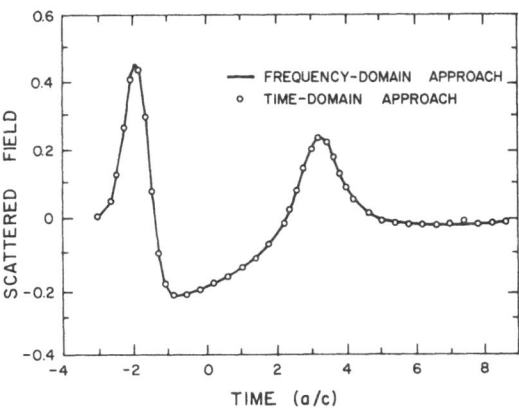

Fig. 2.16. Approximate impulse response of a sphere in the backscatter direction

---

[3] The Mie series solution may be found in a number of classical texts on electromagnetic theory and optics, ..., see, for instance M. BORN and E. WOLF: *Principles of Optics* (Pergamon Press London, New York 1964) Chapt. 13.

the sphere. It is evident that the agreement between the two results is excellent. A similar comparison in the frequency domain results, derived this time by Fourier transforming the numerically derived time-domain results, and from the classical Mie series, is shown in Fig. 2.17. It is clear from the above diagram that the numerical results begin to deviate from the analytical solution above $ka \approx 5$ for the example chosen. The numerical results could be improved by choosing a shorter

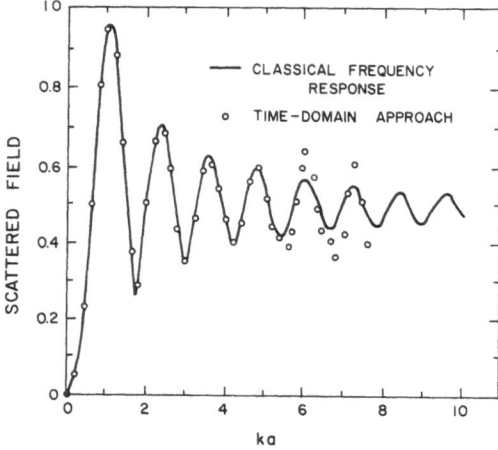

Fig. 2.17. Frequency response in backscatter direction of a sphere of 1-m radius

incident pulse with higher frequency content, though at the cost of increased computational time. Another approach might be to use an analytical extrapolation technique for the high frequency regime as

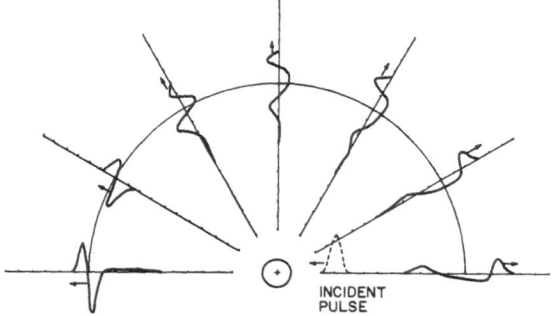

Fig. 2.18. Approximate impulse response of a sphere in E plane

discussed by BENNETT et al. [2.10]. We will briefly mention this technique again in Section 2.7.

The E- and H-plane impulse responses of the sphere illuminated by a Gaussian pulse are exhibited in Figs. 2.18 and 2.19. These figures exhibit the expected creeping wave response (second peak) of the back-scattered direction in the time response.

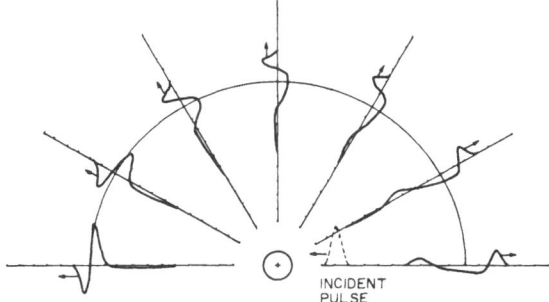

Fig. 2.19. Approximate impulse response of a sphere in the H plane

Another interesting structure is the sphere-capped cylinder whose impulse response for the axial incidence case is shown in Figs. 2.20 and 2.21. This structure has been investigated by BENNETT and WEEKS [2.6] in considerable detail.

The capped cylinder was 1 m in diameter and 3 m in overall length. It was sampled every 30 degrees in azimuth and every 0.25 m along the cylindrical axis of the cylinder proper. In addition, there were three 30-degree-wide bands at each end having 4, 8 and 12 samples in azimuth

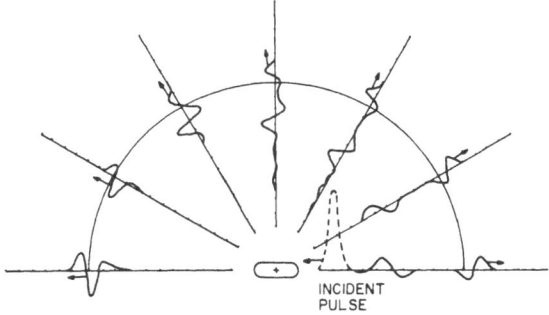

Fig. 2.20. Approximate impulse response of a capped cylinder with axial incidence in the E plane

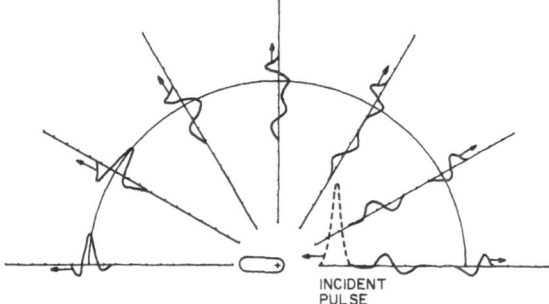

Fig. 2.21. Approximate impulse response of a capped cylinder with axial incidence in the H plane

proceeding from the sphere tip towards the cylindrical section, making a total of 144 patches altogether. Strong indication of creeping wave return is again evident from the presence of the second pulse in the back-scattered return. The widely separated back-scattered returns due to the direct-scatter and creeping-wave mechanisms produce the expected oscillatory frequency response shown in Fig. 2.21. This response is similar to the well-known return from linear dipole antennas.

The last example considered is that of an electrically thin structure— a 20.32 cm (8 in) diameter disk. This structure was analyzed by BENNETT et al. [2.11] using the procedure outlined in Subsection 2.3.3. It may be recalled that their procedure was based on a finite difference type of solution for the EFIE. The backscattered impulse response of the disk is shown in Fig. 2.22 for the case of normal incidence. For this

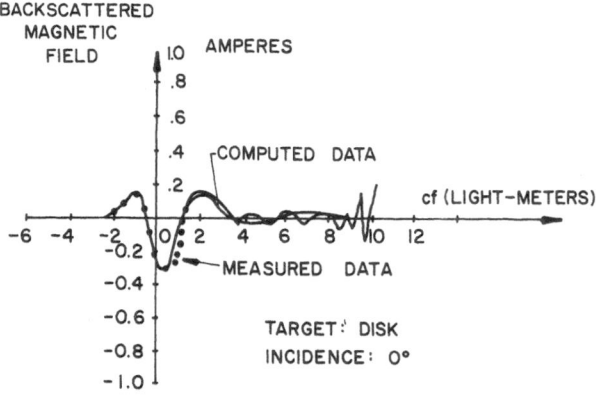

Fig. 2.22. Impulse response of a disk

case, the first portion of the return appears as a smoothed doublet. This is followed approximately 0.8 ns later by a positive pulse which represents the doubly-diffracted wave. Comparison with measured data is also shown in Fig. 2.22 and is found to be generally good. It is conjectured that the instabilities in the computed data are attributable to the manner in which the finite difference technique is applied in the neighborhood of the edge where one of the components of the surface current is singular.

Although we have presented only a few representative results, it is evident that the time-domain response of a scatterer often contains useful information about the nature of scattering, e.g., edge scattering, creeping-wave contribution, in an easily interpretable form. The primary reason for this is that these contributions often are distinctly separated in time allowing them to be identified conveniently. In contrast, this feature is not present in the frequency-domain results derived from a cw analysis, where only an aggregate of various contributions appears in the final result.

Finally, as alluded to several times in this section, it appears that the impulse response from any scatterer may be conveniently expressed in terms of its singularity expansion, which is a summation of complex exponentials. This will be demonstrated in Section 2.6.

2.5 Comparison of Computation Times with Frequency and Time-Domain Approaches

The frequency domain approach for solving transient scattering problems entails, as a first step, the computation of the frequency response of the structure followed by the Fourier transformation of this response. In contrast, the time-domain results are computed directly via the techniques described in this chapter. One might wonder if an advantage is gained in this direct mode of solution over the Fourier transformation approach. Because the time-domain solution for the scattering problem is constructed by marching on time, the solution thus obtained is clearly valid only for the incident wave arriving from a particular direction. In contrast, the system matrix in the frequency domain solution is independent of the incident wave. Thus, if one desires the solution for a large number of incident waves arriving from different directions, one would obviously be better off in following the frequency-domain approach. If, on the other hand, one desires the time-domain response for only the early time period, the frequency domain would still require the computation of the frequency response up to the maximum effective frequency, and the entire range of frequency response would have to be transformed. The time-domain approach will be more efficient in this situation since the interative solution can be conveniently terminated at any given desired time. It is evident that both the computation time as well as storage requirement grow steadily with increasing width of time window to be studied; if this window is small both the computation time and storage requirement would decrease. Even when the final result desired is the complete response, the computation time in the two approaches can be substantially different. This may be seen

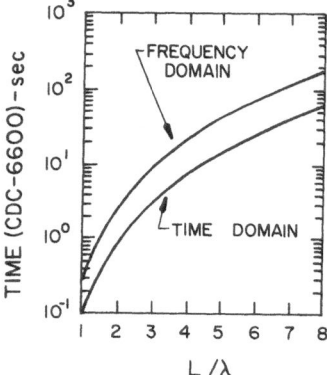

Fig. 2.23. Computer time requirements for the transient analysis of wire structures

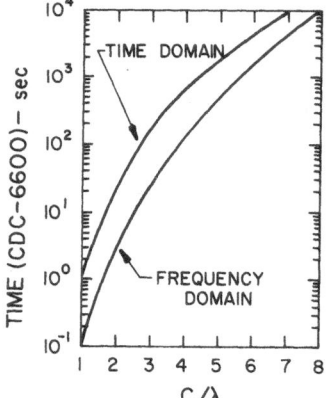

Fig. 2.24. Computer time requirements for the transient analysis of surface structures

by reference to Figs. 2.23 and 2.24, computed by MILLER [2.14], which show that the time-domain approach is more efficient for a wire structure but not for the surface structure. Figures 2.25 and 2.26 show a similar comparison for monostatic-bistatic computations. The last two figures show that the time-domain comutations are relatively more efficient than the frequency-domain method for computing bistatic response, *both* for wire and surface structures.

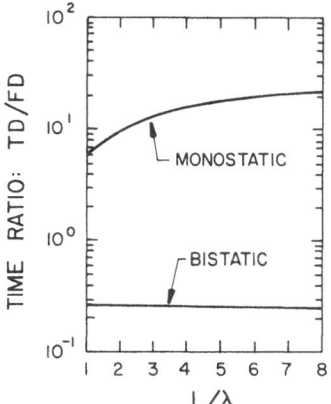

Fig. 2.25. Ratios of time-to frequency-domain calculation times for the transient response of wires

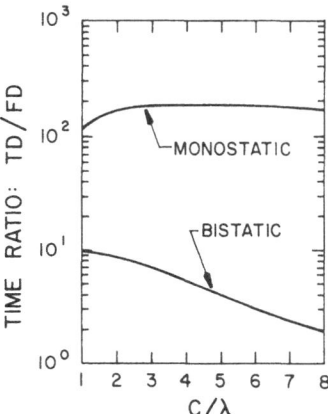

Fig. 2.26. Ratios of time-to frequency-domain calculation times for the transient response of surfaces

Finally, the computer storage requirement is typically much less in the time domain, by a factor of 10 or more for wire structures and 30 or more for surface structures. The storage requirement can be further reduced, in addition to the computation time itself, by using the technique described in the next section.

2.6 Complex-Exponential Representation
of Time-Domain Response

It will be demonstrated in Chapter 3 that under certain circumstances
the transient response of a scatterer may be written in terms of its
singularity expansion which takes the form of a summation[4] of damped exponentials. In this section we show how one can derive such a
representation directly from the impulse response of a scatterer. It has
been pointed out earlier in Sections 2.4 and 2.5 that such a representation is very useful for efficient extrapolation of the time-domain
response that has been computed only for earlier time steps. In addition,
since the singularities depend only on the geometrical characteristics
of the scatterer and not on the incident wave, the SEM (Singularity
Expansion Method) representation is useful in characterizing a target
from its short pulse radar signature. We will now detail the steps for
processing a given impulse response $I(t)$ and for transforming it in
the form

$$I(t)= \sum_{m=1}^{N} A_m e^{s_m t}, \quad t>0.\tag{2.55}$$

Since $I(t)$ is real, it may be readily shown that the complex exponents
(poles in the complex frequency plane) s_m and the weight coefficients
A_m (residues) come in conjugate pairs. However, the method we describe
is able to handle a general complex function $I(t)$.

Since in practice one always deals with a discrete set of sampled
transient data, (2.55) can be rewritten as

$$I(t_n)=I_n= \sum_{m=1}^{N} A_m e^{s_m n \Delta t}= \sum_{m=1}^{N} A_m z_m^n, \quad n=0,1,...,2N-1 \tag{2.56}$$

where

$$z_n = e^{s_m \Delta t} \tag{2.57}$$

and Δt is the size of the time-stepping interval used in obtaining the
sampled data. This set of equations (2.56) may be viewed as M nonlinear

---

[4] Under certain conditions it may be necessarty to include an entire function to this
summation in terms of the damped exponentials. However, in this section we only
consider the case in which the entire function has been eliminated from the representation
by suitable manipulations, e.g., deconvolution of the transient response.

equations for $2N$ unknowns A_m and s_m. An iterative solution of these coupled nonlinear equations is extremely difficult and typically very inaccurate. In the following we discuss a systematic approach—based on Prony's method—that not only circumvents these difficulties but is numerically efficient as well.

The essence of Prony's method is the separation of the problem into that of first finding the exponents (poles) and then finding the residues. Prony has shown that there exists a vector $\{\alpha\}$ with elements $\alpha_1, \alpha_2, \ldots \alpha_N$ such that

$$\sum_{p=0}^{N} \alpha_p I_{p+k} = 0, \quad \text{for} \quad p+k = n = 0,1,2 \ldots M-1 \tag{2.58}$$

Without loss of generality we may set $\alpha_N \equiv 1$ and write (2.58) as

$$\sum_{p=0}^{N-1} \alpha_p I_{p+k} = -I_{N+k}. \tag{2.59}$$

Since the I_{p+k} and I_{N+k} are simply the known sampled transient data values, we may solve this set of equations to obtain the α_p. It should be noted that if $M=2N$ then the solution of this set of equations simply involves the inverse of a real symmetric circulant matrix. If $M>2N$ then the solution can be obtained using a leastsquare approach. Upon substituting $I_{p+k} = I(t_{p+k})$ from (2.56) into (2.58) we may readily derive the equation

$$\sum_{p=0}^{N} \alpha_p z^p = 0 \quad \text{for} \quad z = z_1, \ldots, z_m. \tag{2.60}$$

The roots z_m of this N-th order polynomial equation may be obtained using any polynomial root finding routine. From the roots z_m, the poles s_m are obtained as

$$s_m = \frac{\ln z_m}{\Delta t}. \tag{2.61}$$

It is now a simple procedure to obtain the residues A_m by solving the matrix equation embodied in (2.56) since the elements of the matrix which involve the s_m's are now known. If $M=2N$ then the matrix contained in (2.56) that must be inverted is in the form of a transpose Vandermonde matrix whose inverse can be computed in closed form.

Thus Prony's algorithm simply involves the solution of two matrices and a solution of the zeros of an N-th degree polynominal, N being the number of desired poles. The method requires that in order to find N poles and residues it is necessary to have at least $2N$ equally spaced transient data samples. As pointed out earlier, the realness of the transient response, $I(t)$, requires that the N poles and residues come in complex conjugate pairs.

After the poles s_m and the residues A_m have been determined, it is then possible to express the impulse response of the system for all positive time using (2.55). If, in addition, the frequency domain transfer function is desired, then using the poles and residues, it can be simply expressed as

$$I(-i\omega) = \sum_{m=1}^{N} \frac{A_m}{-\sigma_m - i(\omega - \omega_m)}, \tag{2.62}$$

where poles s_m have been written in terms of their real and imaginary part as

$$s_m = \sigma_m - i\omega_m. \tag{2.63}$$

Thus the frequency domain transfer function can be obtained directly from the time domain impulse response without having to perform a Fourier transform.

To illustrate the application of the technique, we consider the transient behavior of a dipole antenna and derive the complex poles of this structure from its time domain response. The parameters chosen for this example are: dipole length: 1 m (half length/radius) ratio: 100. The induced current at the center of the dipole has been calculated for a Gaussian input pulse $\exp[-A^2(t-t_{max})^2]$ with $A = 5 \times 10^9 \, \text{s}^{-1}$, $t_{max} = 5.556 \times 10^{-10}$ s, and a time step $\Delta t = 5.556 \times 10^{-11}$ s. Although the true impulse response was not used but only the currents produced by a Gaussian excitation, the error introduced was small as the pulse width was quite narrow. The transient response was computed for 500 time steps using the technique outlined in Subsection 2.3.2. The current on one of the center segments is plotted in Fig. 2.27.

In order to calculate the poles the current waveform on one of the center (source) segments was used. Starting at time step number 59 of the calculated current, 150 samples were taken at every third time step. Samples were not taken until time step 59 in order to let the initial driving function, to become negligibly small. A least squares solution was used to obtain 20 complex poles and residues. Figure 2.28 is a plot of the complex pole locations in the second quadrant

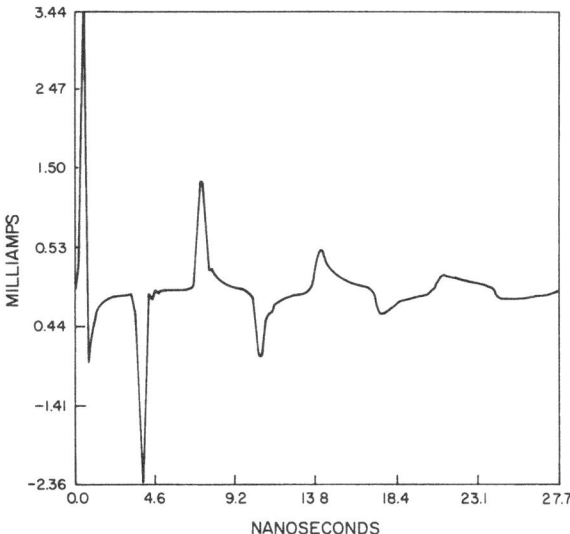

Fig. 2.27. The transient response function for 1.0 meter dipole

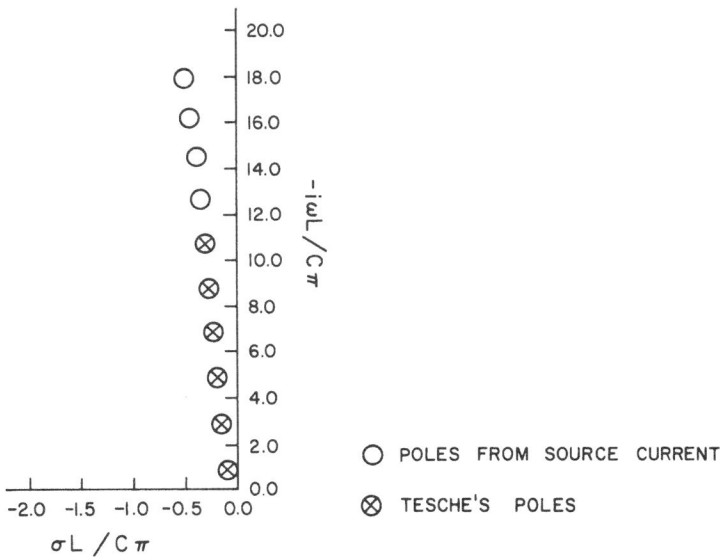

Fig. 2.28. Pole locations in the complex frequency plane for the 1.0 meter dipole

of the complex frequency plane. Only the second quadrant is shown because the 20 poles are actually 10 complex conjugate pairs. The first six of these poles compare well with those calculated by Tesche [2.25] using the frequency domain approach discussed in Chapter 3. The ten pole pairs obtained correspond to the first ten even poles of this dipole. Only the even poles were obtained because the antenna was driven symmetrically. No more than the ten even poles were expected because of knowledge of the original model. Because of the thin wire model and the width of the Gaussian pulse, the expected spectral response has an upper frequency limit of about 10 L/λ. The complex resonances for a dipole occur in the neighborhood of $\lambda/L = 1/2, 3/2, 5/2 \ldots$. Thus, with the upper frequency limit given, only the first ten even resonances can occur.

As a test of the validity of the technique, the input admittance for this dipole was obtained by dividing (2.63) by $V(-i\omega)$, the frequency spectrum of the input Gaussian pulse. The input admittance which is plotted in Figs. 2.29 and 2.30, compares closely for all but the higher frequencies with the input admittance obtained by taking the Fast Fourier Transform of the original data.

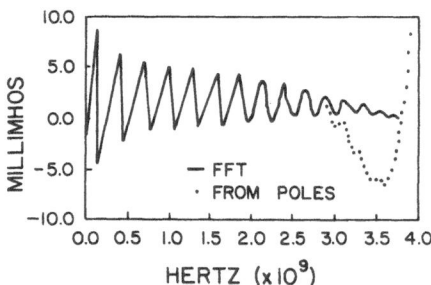

Fig. 2.29. The imaginary part of input admittance for 1.0 meter dipole

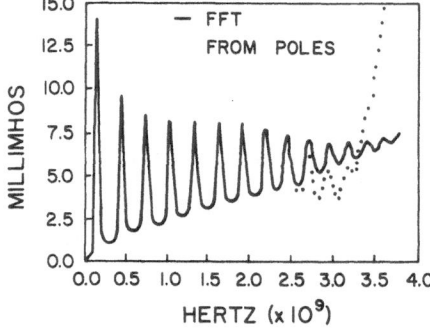

Fig. 2.30. The real part of input admittance for 1.0 meter dipole

It may be of interest to compare the theoretical results with some experimental data which was generated on the transient electromagnetic measurement range at Lawrence Livermore Laboratory [2.26]. The response used here was that of a 1.0 meter monopole located on a ground plane and excited by a Gaussian-pulse plane wave. The monopole was loaded at its base with a 50 Ohm load and the voltage across this load was measured with a sampling scope. A total of 512 samples were taken at a time interval of $\Delta t = 0.4 \times 10^{-10}$ s. Of the 512 measured values only 100 samples at every fifth time step were used. Figure 2.31 shows the measured response in terms of the current through

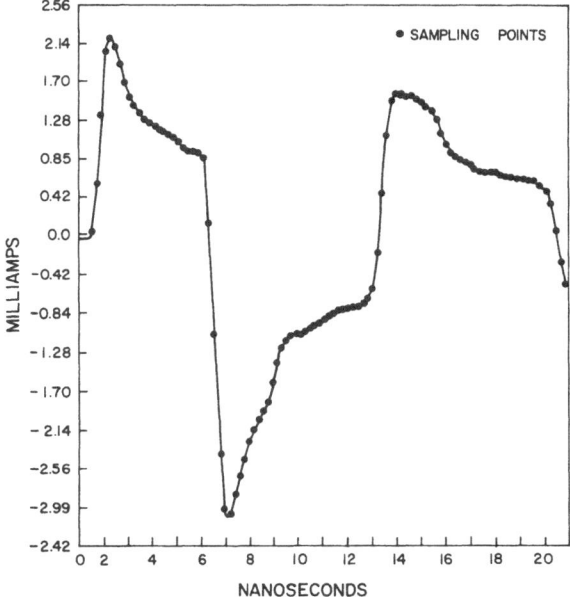

Fig. 2.31. Original experimental data showing the 100 sampling points

the load with the circles showing where the 100 samples were taken. These 100 current samples were used with Prony's method and 41 poles and residues were produced. These 41 poles and residues were then used in (2.56) to reproduce the transient response of the current for 500 time steps where the time step used was five times the original time step. Figure 2.32 shows that his method can be used to extrapolate the measured response to very late time values. Thus with the storage of only 82 complex numbers the original measured time response was

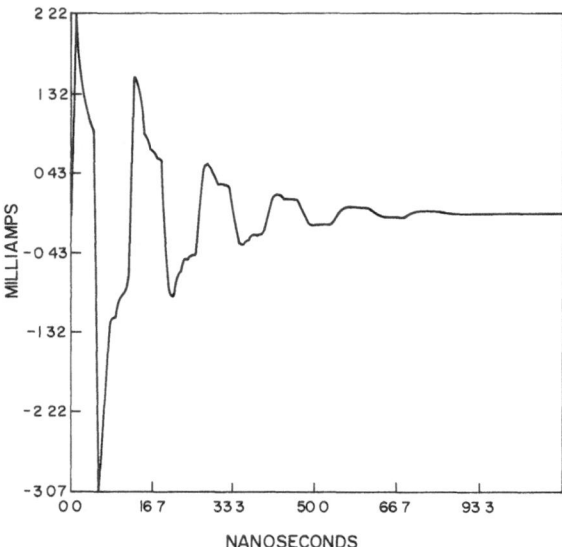

Fig. 2.32. The extrapolated values of the transient response for experimental 1.0 meter whip

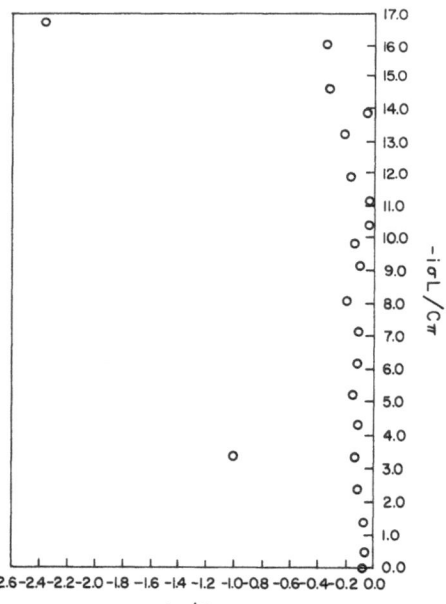

Fig. 2.33. Pole locations in the complex frequency plane for the 1.0 meter experimental whip

reproduced and extrapolated to late values. Figure 2.33 is a plot of the generated poles found in the second quadrant of the complex plane. The first thing that is apparent about these poles is that they tend to fall along a curve running approximately parallel to the imaginary axis. This is typical of the pole locations for a dipole. The frequency of the first nine poles in this layer corresponds to the first nine complex resonant frequencies of a 1-meter monopole. The fact that the remaining poles do not correspond to physical poles again relates to the fact that the Gaussian pulse used did not contain frequencies higher than that of the ninth resonance. The real part of these poles seems to be oscillating around the correct value but this is probably due to the sensitivity of the real part to the noise in the data. The measured data were indeed noisy and no attempt was made to smooth them. Also it is seen that there is a pole sitting on the real axis close to the origin. This is probably because there was a late time dc level present due to the pulser used.

To conclude this section, we observe that the SEM pole singularities of a structure can be derived directly from its time-domain response, regardless of whether it is obtained theoretically or experimentally. That this approach can be employed to substantially reduce the computation time is evident from the discussion presented in this and the previous two sections.

2.7 Extensions of Time-Domain Methods

Having presented a description of the time-domain integral equation technique and a few of the representative results that may be obtained using this technique, we will now briefly discuss possible extensions of the various approaches that have been outlined in the previous sections. Specifically, we consider the questions of extension of time-domain methods to complex strcutures and to short-pulse excitations that have large bandwidths. We also touch on the problems of increasing the efficiency of computation and explore the possibility of handling nonlinear effects.

Although we have only reported the results for simple wire or surface structures, several complicated shapes such as satellite structures with protruding wire appendages have in fact been studied by a number of authors including BENNETT et al. [2.27] and BENNETT and WEEKS [2.28]. One possible approach to complex geometries is to model them with thin wires. This approach, which has been followed quite extensively (see [2.7]) in the frequency domain for a wide variety of structures, has also been found useful in the time-domain. One advantage of this approach is that it allows a convenient extension of the thin-wire computer code that may already be available. Of course, certain modifications are necessary in the thin-wire approach described in Subsection 2.3.2 so that the program is able to accommodate multiple-wire junctions. This program requires some sophisticated interpolation of the currents in the various arms of the junction as they flow past the junction point. Techniques for this have been described, among others, by VAN BLARICUM [2.20], in considerable detail and the interested reader is referred to this work for additional information. Some frequency domain approaches that have been developed by CHAO and STRAIT [2.29], and MITTRA and KO [2.21] may also be adaptable in the time domain and, if so, would require considerably less modification of the conventional thin-wire programs than the interpolation technique mentioned above.

Another class of configurations of practical interest may be described as hybrid structures, which are combinations of surface and wire geometries, e.g., spheres and cylinders with wire appendages that might represent models of satellites. These structures could be analyzed by combining the programs for the general three-dimensional surface and one-dimensional wire structures described in Subsections 2.3.1 and 2.3.2. It is, of course, possible to treat these structures using the EFIE alone, which can be employed for both the wire and surface-type structures. However, as pointed out in Section 2.3, the MFIE equation is more efficient for surface structures and its use is preferred whenever possible. The EFIE approach does, however, yield a general purpose program which requires less bookkeeping.

The last category of complex structures we consider is that with shell-like geometries. It has been found that care must be exercised in handling shell structures if one is to avoid numerical instabilities that arise because of the singular behavior of the current in the neighborhood of sharp edges and corners. The instability problem has been investigated in some detail by DAVIS and MITTRA [2.19], in the frequency domain but the general conclusions apply to the time-domain also. One possible method (see [2.19] for details) for circumventing the problem caused by the presence of edges is to use a smooth, spline function representation for the finite part of the edge current and to incorporate the known edge-behavior of the current for the singular part of the current in the vicinity of the edge. An alternative approach is to use careful interpolation of the current and the vector potential in the neighborhood of the edge, as has been described by BENNETT et al. [2.11]. More effort could be directed toward improving these techniques so that the structures with edges could be handled without excessive complications.

Yet another desirable extension of the time-domain integral-equation methods would be to increase their capability of handling body sizes that are large compared to the effective bandwidth of the incident pulse. Even though no matrix inversion is involved in the time-stepping approach, the storage of the past-time information becomes prohibitive when dealing with large structures. One is typically limited to body sizes whose characteristic dimensions are less than 1λ, in terms of the maximum effective frequency of the input pulse.

In principle, of course, one can solve the deconvolution problem to obtain the impulse response of a scatterer by using the formula

$$A(\omega) = \frac{F(\omega)}{S(\omega)}, \tag{2.64}$$

where $F(\omega)$ and $S(\omega)$ are the Fourier transforms of the time-domain waveforms of the scattered and input fields, respectively. The desired time-domain impulse response is obtained by Fourier transforming $A(\omega)$. however, as is typical in the deconvolution of this type, the computational noise that is inherently present in the numerical data sets an upper bound of the frequency for which accurate deconvolution can be achieved, and of the $A(\omega)$ that can be derived from the use of (2.64) with relatively little error. As an example of such error-buildup we show in Figs. 2.29 and 2.30 the results of imaginary and real parts of the frequency-domain response for the input admittance of an antenna of 1 m length which has been driven by a Gaussian pulse $\exp(-At^2)$ with $A = 3.25 \times 10^9$. These results, which have been derived via the use of the deconvolution technique just described, clearly show the effect of serious error-buildup due to computation noise above $L/\lambda \approx 9$, which is consistent with the parameters for the problem. BENNETT et al. [2.10] have proposed an extrapolation technique which involves the following two steps. In their method a smoothing operation is first applied to the deconvoluted data in order to suppress the noisy behavior of these data beyond the effective cutoff frequency of the input pulse. An extrapolation technique, based on asymptotic methods such as the geometrical theory of diffraction, is subsequently applied to the smoothed data to extend the response to higher frequencies. Care must be exercised in 'patching up' the low and high frequency responses so that the resultant composite response will, when transformed in the time-domain, produce results that satisfy the causality condition. This is not as easily accomplished as might appear at first sight, since the causality condition does not have a convenient counterpart in the frequency domain.

It might be recalled that another possibility that appears promising for alleviating the storage problem is to employ the complex exponential representation discussed in Section 2.6. This approach allows convenient and efficient storage of the time-domain data in terms of a relatively few complex poles and residues, thus reducing the storage requirements rather substantially.

Another desirable extension of the solution techniques for the conventional time-domain integral equation would be to increase the computational efficiency of these methods. Several approaches have been proposed for doing this, the principal one being use of adaptive time stepping. It is a common feature of the time-domain response functions that most of the rapid variations occur in the early-time, while the later-time response contains only a few of the decaying exponentials, and thus has relatively smooth variations. For this reason it would be desirable to use shorter-time steps initially, and then increase the step-size during the late-time periods where only the low frequency content of the spectrum dominates the behavior. This method has in fact been successfully applied to two-dimensional time domain problems by BENNETT and WEEKS [2.6], but has not been extensively used. Yet another approach, which has been mentioned earlier, would entail the use of SEM type representations derived from the early-time data and their subsequent extrapolation into later-time periods. Further exploration into these techniques would indeed be useful.

2.8 Applications

There are a number of applications of the time-domain analysis presented in this chapter, some of which are quoted below. Much of the impetus for the work on time-domain integral equations by BENNETT and his colleagues came from a program for object identification using short pulse radar. Transient analysis of antennas is also useful for synthesis of pulse shapes with applications in sensing and simulation, for example, in EMP studies. Much work in this area, which has been carried out under the aegis of a program sponsored by the Air Force Weapons Laboratory, Kirtland Air Force Base, Albuquerque, New Mexico (USA) has been described in the extensive series of EMP notes (see Chapter 3) published by the above organization.

Another area particularly suited to transient analysis is the investigation of antennas or scatterers that are actively loaded with nonlinear elements. Using the pole extraction technique described in Section 2.6 it is possible to construct (directly) an equivalent lumped circuit for an antenna from its time-domain response. This circuit can be conveniently modified to take into account the effect of active loading, and the composite circuit can be subsequently investigated by using one of the available programs for analyzing such nonlinear but lumped networks. It goes without saying that the knowledge of the time-domain data is essential for constructing the lumped parameter equivalent network for the structure and that such a circuit cannot be conveniently obtained from a frequency domain analysis of the same structure.

2.9 Conclusions

We have made an attempt in this chapter to present an overview of some of the representative time-domain integral-equation approaches for solving a wide class of transient electromagnetic radiation and scattering problems. We have compared the time-domain and frequency-domain approaches and have pointed out their relative advantages and disadvantages. We have indicated certain areas where further work would be useful and have described some unique applications of time-domain techniques. It is hoped that the reader will be able to gain a broad perspective of the various time-domain techniques and become aware of the scope and limitation of some of these methods.

Acknowledgements

The author is pleased to acknowledge extensive help and cooperation he received from many individuals during the course of preparing the manuscript. Notable among these are Dr. C. L. BENNETT of Sperry Rand Corporation, Dr. E. K. MILLER of the Lawrence Livermore Laboratory, and Mr. M. L. VAN BLARICUM of the Electromagnetics Laboratory of the University of Illinois. Comments from Dr. C. BAUM and the editor (L. B. FELSEN) were extremely helpful in improving the manuscript.

The author wishes to thank all of these individuals for their kind assistance, patience and encouragement.

References

2.1   R. MITTRA (ed.): *Computer Techniques for Electromagnetics* (Pergamon Press, New York 1973)
2.2   E. M. KENNAUGH, R. L. COSGRIFF: "The Use of Impulse Response in Electromagnetic Scattering Problems", in: IRE Natl. Conv. Rec. 1958, pt. 1, p. 72
2.3   E. M. KENNAUGH, D. L. MOFFATT: Proc. IEE **53**, 893 (1965)
2.4   W. L. WEEKS: "Random Signal Radar"; Purdue University School of Electrical Engineering, Lafayette, Indiana, Final Technical Report 4696 (June, 1967)

2.5 J. RHEINSTEIN: IEEE Trans AP-**16**, 89 (1968)

2.6 C. L. BENNETT, W. WEEKS: "A Technique for Computing Approximate Electromagnetic Impulse Response of Conducting Bodies"; Purdue University, Lafayette, Indiana, Technical Report TR-EE68-11 (1968)

2.7 A. J. POGGIO, E. K. MILLER: Reference 2.1, Chapt. 4

2.8 E. P. SAYRE, R. F. HARRINGTON: "Transient Response of Straight Wire Scatterers and Antennas", in: Proc. Intern. Ant. Prop. Symposium, Boston, Mass. 1968, p. 160

2.9 A. M. AUCKENTHALER, C. L. BENNETT: IEEE Trans. MTT-**19**, 892 (1971)

2.10 C. L. BENNETT, A. M. AUCKENTHALER, R. S. SMITH, J. D. DELORENZO: "Space-Time Integral Equation Approach to the Large Body Scattering Problem", Final Report, F30602-71-C-0162, RADC-CR-73-70, AD 763 794, May, 1973

2.11 C. L. BENNETT, K. S. MENGER, R. HIERONYMUS: "Space-Time Integral Equation Approach for Targets with Edges", Final Report F30602-73-C-0124, SCRC-CR74-3, March 1974

2.12 E. K. MILLER, A. J. POGGIO, G. J. BURKE: "An Integro-Differential Equation Technique for the Time-Domain Analysis of Thin Wire Structures; Part I, The Numerical Method", Lawrence Livermore Laboratory, Report UCRL73346 (1971)

2.13 A. J. POGGIO, E. K. MILLER, G. J. BURKE: "An Integro-Differential Equation Technique for the Time-Domain Analysis of Tin Wire Structures; Part II, The Numerical Results", Lawrence Livermore Laboratory, Report UCRL-73346 (1972)

2.14 E. K. MILLER: "Some Computational Aspects of Transient Electromagnetics", Lawrence Livermore Laboratory, Report UCRL-51276 (1972)

2.15 D. E. MEREWETHER: IEEE-Trans. EMC-**13**, 41 (1971)

2.16 O. D. KELLOG: *Foundations of Potential Theory* (Dover Publications, New York 1953)

2.17 R. MITTRA, Y. RAHMAT-SAMII, D. V. JAMNEJAD, W. A. DAVIS: Radio Science **8**, 869 (1973)

2.18 R. F. HARRINGTON: *Field Computations by Moment Methods* (MacMillan, New York 1968)

2.19 W. A. DAVIS, R. MITTRA: "Numerical Solutions to the Problems of Electromagnetic Radiation and Scattering by a Finite Hollow Cylinder", University of Illinois, Urbana, Illinois, Technical Report 74-10 (October 1974)

2.20 M. VAN BLARICUM: "A Numerical Technique for the Time-Dependent Solution of Thin-Wire Structures with Multiple Junctions", M. S. Thesis, Electrical Engineering Department, University of Illinois (1972)

2.21 W. L. KO, R. MITTRA: "A Finite Difference Approach to the Wire Junction Problem", IEEE Trans. AP-**23**, 435 (1975)

2.22 M. L. VAN BLARICUM, R. MITTRA: "A Technique for Extracting the Poles and Residues of a System Directly from Its Transient Response", IEEE Trans. AP-**23**, 777 (1975)

2.23 J. W. COOLEY: "Finite Complex Fourier Transform", IBM SHARE Program Library No. SDA 3465, New York (1966)

2.24 R. PRONY: "Essai experimental et analytique, etc.", Paris J. l'Ecole Poltechnique, 1, cahier 2, 24–76 (1975)

2.25 F. M. TESCHE: IEEE Trans. AP-**21**, 53 (1973)

2.26 R. A. ANDERSON, J. A. LANDT, F. J. DEADRICH, E. K. MILLER: "The LLL Transient Electromagnetic Measurement Facility: A Brief Description", UCID-16573, Lawrence Livermore Laboratory, Livermore, California (August 9, 1974)

2.27 C. L. BENNETT, J. D. DE LORENZO, A. M. AUCKENTHALER: "Integral Equation Approach to Wideband Inverse Scattering", Rome Air Development Center, Rome, New York, Report RADC-TR-70-177, Vol. I (1970)

2.28 C. L. BENNETT, W. L. WEEKS: IEEE-Trans. AP-**18**, 627 (1970)

2.29 H. H. CHAO, B. J. STRAIT: IEEE-Trans. AP-**19**, 701 (1971)

3. The Singularity Expansion Method

C. E. BAUM

With 26 Figures

This chapter considers a procedure for respresenting and calculating efficiently the transient electromagnetic response of antennas and scatterers, as well as of other types of electromagnetic configurations. The present considerations, however, are limited to antennas and scatterers. The method is also useful for representing and calculating efficiently the broadband frequency response characteristics as one might expect from the Laplace (Fourier) transform relations connecting the time domain and frequency domain. The method, called the singularity expansion method (SEM), is based on the analytic properties of the electromagnetic response as a function of the two-sided Laplace transform variable s which can be called the complex frequency. The singularities of the Laplace transform are used to characterize the electromagnetic response in both the time domain and complex frequency domain. Application of the method to electromagnetic scattering problems is rather new and has been studied extensively only for objects in free space. Indications are that SEM can be used profitably in other types of situations, but this is a matter for further investigation. This chapter develops the SEM approach for objects in free space, considers some specific examples, and indicates some of the directions of future research.

Most of the literature on the application of SEM to electromagnetic antennas and scatterers is in the form of internal but widely distributed reports. These reports are found in the note series on EMP (electromagnetic pulse) and related subjects which are edited and distributed by this author (e.g. [3.1,2]). Notes relevant to SEM are found in the Interaction Notes, Sensor and Simulation Notes, Theoretical Notes, and Mathematics Notes[1]. However, some of these reports have recently appeared in the open literature [3.3–5].

---

[1] Available from C. E. BAUM, Air Force Weapons Laboratory (EL), Kirtland AFB, New Mexico 87117 or Defense Documentation Center, Cameron Station, Alexandria, Va. 22314.

3.1 Background Leading to the Development of the Singularity Expansion Method (SEM)

3.1.1 Experimental Observations

Development of the singularity expansion method was stimulated by observation of the general characteristics of typical transient responses in experiments on various complicated scatterers (e.g., aircraft) [3.2]. Transient response waveforms of such objects appear to be dominated by a few damped sinusoids. These are not the only types of functions appearing in such transient data but they are quite prevalent, especially for long observation times. This may be confirmed from a study of transient response data recorded in various examples throughout this volume.

The above-mentioned experimental observations were for broadband transient electromagnetic excitation, wherein "all" frequencies were present. Since the Laplace transform of a damped sinusoid corresponds to one pole or a pair of poles in the complex frequency plane, the scattering object may be expected to have a large response at frequencies near such poles. A broadband pulse excites these poles which can be referred to as the natural frequencies of the object. Such natural frequencies have corresponding natural modes which are field, charge or current solutions of the "free oscillation" problem; the modal distributions are not functions of the incident field or other aspects of the source configuration. The complex amplitude coefficients of the natural modes are dependent on the source function; such coefficients, which can be regarded as the strengths of the "resonances", will later on be referred to as "coupling coefficients".

Having identified damped oscillatory terms in the transient response of an object as poles in the complex frequency plane one can ask what part of the complete solution such terms represent. This question leads to the singularity expansion method.

3.1.2 Basic Aspects of SEM

Consider the solution of the electromagnetic problem as a function of the complex frequency s, where s is the Laplace transform variable corresponding to the time variable t. If $f(t)$ is some arbitrary time function, then we define the Laplace transform (two-sided) with symbol $\sim$ as

$$\tilde{f}(s) \equiv \int_{-\infty}^{\infty} f(t) e^{-st} dt . \tag{3.1}$$

This definition applies to vector and dyadic functions as well as scalar functions. Note that by setting $s = -i\omega$ the above definition gives a Fourier transform corresponding to the time harmonic convention used elsewhere in this book.

With suitable boundedness requirements on $f(t)$, including asymptotic bounds for $t \to \pm\infty$, the Laplace transform is an analytic function of s in the complex variable sense, at least for portions of the s plane. Analytic continuation extends the domain of analyticity to other portions of the s plane. For various values of s the Laplace transform will not be an analytic function of s; such values of s are termed "singularities" and their utilization leads to the name "singularity expansion method" (SEM). By understanding the properties of such singularities, including the singularities as $s \to \infty$ (entire function), one can characterize $\tilde{f}(s)$ everywhere in the s plane. Various texts such as [3.40, 43, 46] consider the general properties of functions of a complex variable. SEM uses the properties of the solution (explicit or implicit, closed form or numerical) as an analytic function of the complex frequency to represent the solution of electromagnetic problems[3.2, 3, 5]

Some typical singularities and the inversion contour are illustrated in Fig. 3.1. In order to recover $f(t)$ one may use a contour integral over the Bromwich contour C_0 defined by $\mathrm{Re}\{s\} = \Omega_0$ as

$$f(t) = \frac{1}{2\pi i} \int_{C_0} \tilde{f}(s) e^{st}\, ds = \frac{1}{2\pi i} \int_{\Omega_0 - i\infty}^{\Omega_0 + i\infty} \tilde{f}(s) e^{st}\, ds. \tag{3.2}$$

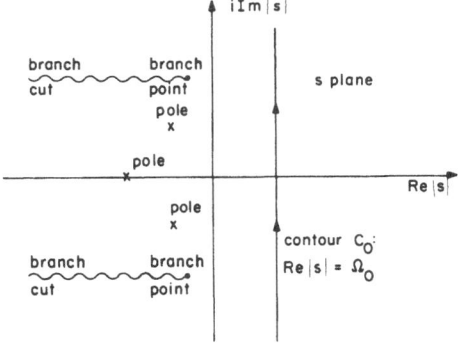

Fig. 3.1. s plane with some singularities and the inversion contour C_0

Using the well known method of contour deformation, one may move into the left half plane to form separate contours around the various singularities and hence define separate terms for $f(t)$, one for each singularity. In so deforming the contour there is, in general, a contour integral at $|s| \to \infty$ as one of the contributions; this is related to the entire function (singularity at ∞). Other contributions correspond to localized singularities such as poles; in such cases the contour integrals are around isolated points in the s plane. Still other terms correspond to branch point singularities from which branch cuts are defined to keep $\tilde{f}(s)$ single valued; these branch cuts may sometimes be contained in the finite s plane by connecting more than one branch point; alternatively and more generally, branch cuts may extend to ∞ in the s plane. The transient response of the object may be characterized by utilizing the properties of the singularity contributions in various ways. This aspect, fundamental to the SEM, is elucidated by the presentation to follow.

A useful aspect of the SEM representation of the electromagnetic solution is its dual form, one for the s plane (time harmonic) and one for the time domain. For the important pole terms, the s domain and t domain forms are both elementary (poles and exponentials). Furthermore the same numerical quantities are calculated for both s and t domain solutions; one set of numbers solves both problems. In that sense one is simultaneously calculating (or measuring) both frequency and time domain parameters (see also Subsect. 1.4.1).

In this chapter we consider passive objects, whose singularities lie in the left half plane. If any singularities lie on the imaginary axis their properties are constrained; e.g., poles with $\mathrm{Re}\{s\} = 0$ can only be simple poles. This requirement means that the delta function (impulse) response of the object decays, or at least does not grow unboundedly, with increasing time (see also Chapt. 1). These fundamental concepts can be applied as well to active, but stable, objects such as active antennas and scatterers. As will be briefly discussed later there are synthesis applications of SEM where the impedance properties of an object are important. For passive objects the impedance properties impose constraints much as they do in passive electrical circuits. This puts additional restrictions on the characterization of the object response in the s plane.

At the present stage of development, it appears that SEM is not so much an alternative method for solving boundary value problems but one which should be used in conjunction with other methods to cast the solution for broadband frequency and time response in a more useful form. The method requires finding the poles, branch points, essential singularities, and entire functions, which characterize a given object. This may be done in various ways, both analytically (with suitable approximations) and numerically. In this connection, it should be noted that certain approximate solutions in the frequency domain use high frequency asymptotic forms which may not be strictly convergent (see Subsect. 1.4.1). Such forms may therefore not be able to provide certain of the singularities correctly, in particular those located in a region of the complex s plane where the approximation is invalid.

The procedures employed in the following applications of SEM to electromagnetic problems may be categorized as analytical, numerical, and hybrid (a combination of both). Analytical methods include but are not necessarily limited to eigenfunction expansions by separation of variables in special coordinate systems, special conformal transformation techniques, special Taylor series expansions, approximate closed form solutions of one-dimensional integral equations for thin-wire structures, asymptotic high-frequency or early-time expansions such as the geometrical theory of diffraction (GTD) (see Chapt. 1). Numerical methods include finite difference forms of the differential equations and boundary conditions, iterative or relaxation methods, and moment methods involving choices of expansion and testing functions over appropriate regions such as boundary surfaces [3.47] (see Chapt. 2). Hybrid methods combine analytical and numerical methods for various parts of the problem with appropriate attention paid to properly matching them together.

3.1.3 Previous Work Related to SEM

The natural frequencies of objects have been discussed occasionally in the electromagnetic literature. These natural frequencies have been referred to as the free oscillations and are those frequencies for which the object has a response without a forcing function. Associated with such natural frequencies are natural modes. Near a natural frequency or resonance the forced oscillations due to sinusoidal excitation have also been considered. In such a situation the nearby natural frequency and mode dominate the response.

Studies of this type have been conducted by various investigators including POCKLINGTON [3.34], ABRAHAM [3.35], RAYLEIGH [3.36], OSEEN [3.37], HALLÉN [3.38], PAGE and ADAMS [3.39], and STRATTON [3.41]. Such studies were greatly facilitated by consideration of thin objects or objects for which a separation-of-variables decomposition of the solution was appropriate. Natural frequencies could be found in exact, but more often approximate, explicit form. For thin-wire objects the above results have been reviewed and extended by MARIN [3.19].

A general theory of the interior or cavity problem has been worked out in terms of natural frequencies and natural modes as in SCHWINGER [3.33] or VAN BLADEL [3.44]. For interior problems it is found that the natural modes form a complete set and have convenient orthogonality properties. When considered from a complex frequency viewpoint this formulation of the interior problem is interpretable as the SEM representation of the interior problem. The cavity modes correspond to first-order poles with $\text{Re}\{s\} = 0$ (in the absence of losses), and can be thought of as a special case of the general SEM. Since SEM includes both exterior and interior natural frequencies the general formulas for poles discussed later in this chapter apply also to the cavity problem.

Sometimes antenna impedances are approximated by lumped element circuits as, for example, by SCHELKUNOFF and FRIIS [3.42]. SEM considerations lead naturally to such a viewpoint in that poles and zeros of an impedance function can be used in synthesizing circuits.

An analogy can also be drawn between the complex frequency for SEM problems and the complex energy, complex angular momentum, etc., in particle physics (quantum mechanics). The analogue of SEM in antennas, scattering, propagation, etc., is a procedure using the dispersion relations [3.45] in particle physics. Mathematical developments concerning completeness, exceptional cases, etc., are very important for the applicability and limitations of SEM. In this connection, the reader is referred to some general observations made by WEINSTEIN [3.48].

3.2 SEM for Various Types of Problems

3.2.1 General Properties of the Response in the Complex Frequency Domain

There are some important properties of the response in the complex frequency domain which are of a quite general nature. First note the property of conjugate symmetry. Simply stated if $f(t)$ is a real-valued time function then

$$\tilde{f}(s^*) = [\tilde{f}(s)]^*, \tag{3.3}$$

where the asterisk denotes the complex conjugate. This property extends to vector and dyadic functions as well. From this one immediately infers that the singularities of $\tilde{f}(s)$ are symmetrically placed with respect to the $\text{Re}\{s\}$ axis. As an example note the singularity placement in Fig. 3.1. Except for the singularities with $\text{Im}\{s\} = 0$ this property reduces the singularities of interest by half. For convenience the upper half plane is used.

The concept of conjugate symmetry can be generalized somewhat to include combined quantities as well. Consider Maxwell's equations for free space:

$$\nabla \times E = -\frac{\partial}{\partial t} B - J_m, \qquad \nabla \times H = \frac{\partial}{\partial t} D + J$$

$$\nabla \cdot B = \varrho_m, \qquad\qquad \nabla \cdot D = \varrho$$

$$\nabla \cdot J = -\frac{\partial}{\partial t} \varrho, \qquad \nabla \cdot J_m = -\frac{\partial}{\partial t} \varrho_m \tag{3.4}$$

$$D = \varepsilon_0 E, \qquad\qquad B = \mu_0 H$$

where both electric and magnetic current and charge densities are included for convenience. Defining

$$Z_0 \equiv \sqrt{\frac{\mu_0}{\varepsilon_0}}, \qquad c \equiv \frac{1}{\sqrt{\mu_0 \varepsilon_0}} \tag{3.5}$$

one may introduce combined fields, current densities, and charge densities as

$$F_q \equiv E + q i Z_0 H$$

$$K_q \equiv J + q \frac{i}{Z_0} J_m \tag{3.6}$$

$$Q_q \equiv \varrho + q \frac{i}{Z_0} \varrho_m$$

with the separation index

$$q = \pm 1 \tag{3.7}$$

so that both choices of q are preserved for our use. The Maxwell equations for these combined quantities reduce to

$$\left[\nabla \times -q \frac{i}{c} \frac{\partial}{\partial t} \right] F_q = q i Z_0 K_q$$

$$\nabla \cdot F_q = \frac{1}{\varepsilon_0} Q_q \tag{3.8}$$

$$\nabla \cdot K_q = -\frac{\partial}{\partial t} Q_q.$$

The combined quantities can be extended to include potentials, natural modes, etc., for the 3-vector electromagnetic quantities [3.26].

While use of combined quantities of this kind has not been very common and has led to some new theorems [3.30] this chapter will not dwell on them. It is expected, however, that this formulation could have significant future impact on SEM and eigenmode representations [3.21].

If one uses a combined formulation then, in the time domain, $f_q(t)$ is complex. Note that we have [3.30]

$$[f_q(t)]^* = f_{-q}(t)$$
$$\tilde{f}_q(s^*) = [\tilde{f}_{-q}(s)]^*$$

$$(3.9)$$

where $f_q(t)$ represents any scalar, vector, or dyadic combined quantity as above, combining two real-valued time functions. This is conjugate symmetry generalized to combined quantities. One might call this combined conjugate symmetry.

Since the electric and magnetic fields are constrained in their relation to each other, then if $\tilde{E}$ has a singularity at some particular s, so does $\tilde{H}$, in general, and conversely. Hence $\tilde{F}_q$ has the same singularities as well. Note that the commonality is in the locations, not the magnitudes, of the singularities. Hence combined quantities have a conjugate symmetric singularity pattern in the s plane such as shown in Fig. 3.1.

3.2.2 Integral Equations

Consider some body with finite linear dimensions in free space, as illustrated in Fig. 3.2. Let this body have a surface S which bounds a volume V. One can write an integral equation for the response of the object in terms of the current density as

$$\int_V \tilde{\Gamma}(r, r'; s) \cdot \tilde{J}(r', s) dV' = \tilde{I}'(r, s) \qquad r, r' \in V,$$

$$(3.10)$$

where $\tilde{I}'$ is some forcing function such as an incident electric or magnetic field, field at an antenna gap, etc., and where $\tilde{\Gamma}$ is the dyadic kernel of the integral equation. The type of integral equation is not specified here since it does not matter for our formal manipulations. The reader is referred to Chapter 2 for more information on integral equations. Note that frequency-domain integral equations, specifically those generalized to complex-frequency domains, are used exclusively in this chapter.

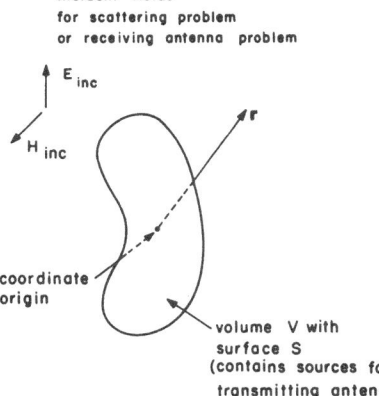

incident fields
for scattering problem
or receiving antenna problem

coordinate
origin

volume V with
surface S
(contains sources for
transmitting antenna
problem)

Fig. 3.2. Radiation and scattering
of electromagnetic fields by an object

For convenience the integral equation (3.10) can be written using a symmetric product notation as

$$\langle \tilde{\underline{\Gamma}}(r,r';s); \tilde{J}(r',s)\rangle = \tilde{I}'(r,s),\tag{3.11}$$

where $\langle\,,\,\rangle$ means to "integrate over the common space coordinates" (common between terms separated by the comma). The domain of integration can be V or S, as required for the integral equation. Symbols above the comma are used to indicate the type of multiplication such as dot product (as in (3.11)), cross product (as in (3.50)), convolution (as perhaps with two functions of time), etc. The unknown current term in (3.11) can be a volume current density, a surface current density, or a line current such as would be used in a thin-wire approximation.

If the Laplace transform of the incident wave or source field $\tilde{I}'(r,s)$ is replaced by the Laplace transform of an appropriately normalized (dimensionless) temporal-delta-function incident wave or source field $\tilde{I}(r,s)$, then one can write an equation for the normalized delta function response as

$$\langle \tilde{\underline{\Gamma}}(r,r';s); \tilde{U}(r',s)\rangle = \tilde{I}(r,s).\tag{3.12}$$

This equation is the one most convenient for developing the general SEM equations.

If the object of concern is complicated then it may be necessary to solve (3.12) numerically. Using the method of moments[2] [3.47] an integral equation such as (3.12) can be converted into a matrix equation in the form (the dependence on r and r' is not shown)

$$(\tilde{\Gamma}_{n,m}(s)) \cdot (\tilde{U}_n(s)) = (\tilde{I}_n(s)) \qquad n, m = 1, 2, 3, \ldots, N \qquad (3.13)$$

thus converting the operator into an $N \times N$ matrix relating N component response and excitation vectors (numerical vectors). The basis and expansion functions are typically arrived at by zoning the body into small sections and using the currents (and their derivatives, etc.) as unknowns, and similarly for the excitation. This chapter is not concerned with the details of MoM but the form of (3.13) is useful for our purposes. In particular it is noted that the elements of $(\tilde{\Gamma}_{n,m}(s))$ are analytic functions of s, being simply related to the dyadic Green's function; this has implications concerning the analytic properties of the response.

3.2.3 Finite-Size Bodies: Poles in the Finite s Plane plus an Entire Function

It has been shown that for certain types of objects, namely finite-size objects in free space consisting of perfect conductors and/or passive media with constitutive parameters suitably constrained in their s plane properties, the object response has only poles as singularities in the finite s plane [3.2, 3, 7]. This is an extremely important property in that the general form of the singularity expansion for such objects is considerably simplified, ruling out both essential singularities and branch point singularities with their associated branch cuts. Such objects have been considered by both the method of moments and integral equation theory.

The various integral equations and corresponding moment representations, (3.12) and (3.13), use various forms of the free-space Green's function. The scalar and dyadic forms satisfy ([3.26, 49] see also Subsect. 1.2.3)

$$(\nabla^2 - \gamma^2)\tilde{G}_0(r, r'; s) = -\delta(r - r')$$
$$(\nabla \times \nabla \times + \gamma^2)\tilde{\underset{\sim}{G}}_0(r, r'; s) = \delta(r - r')\underset{\sim}{1} \qquad (3.14)$$

where $\underset{\sim}{1}$ is the identity dyadic and $\delta(r - r')$ is the delta function (3 dimensional). The associated radiation conditions (for $\text{Re}\{s\} \geq 0$) are

---

[2] "Method of moments" will be abbreviated as MoM. For a recent treatment of this subject see the 3rd volume of this series.

$$\lim_{r \to \infty} r \left(\frac{\partial}{\partial r} + \gamma \right) \tilde{G}_0(r, r'; s) = 0$$

$$\lim_{r \to \infty} r (\nabla \times + \gamma r_0 \times) \boldsymbol{G}_0(r, r'; s) = \boldsymbol{0}, \qquad r_0 \equiv \frac{r}{r}$$

(3.15)

where the complex propagation constant is

$$\gamma \equiv \frac{s}{c}.$$

(3.16)

Defining

$$\zeta \equiv \gamma |r - r'|$$

(3.17)

explicit representations are

$$\tilde{G}_0(r, r'; s) = \gamma \frac{e^{-\zeta}}{4\pi\zeta}$$

$$\nabla \tilde{G}_0(r, r'; s) = \boldsymbol{R}_0 \frac{\gamma^2}{4\pi} (-\zeta^{-2} - \zeta^{-1}) e^{-\zeta}$$

$$\tilde{\boldsymbol{G}}_0(r, r'; s) = \left(\boldsymbol{1} - \frac{1}{\gamma^2} \nabla\nabla \times \right) [\tilde{G}_0(r, r'; s) \boldsymbol{1}]$$

$$= \left(\underset{\sim}{\boldsymbol{1}} - \frac{1}{\gamma^2} \nabla\nabla \right) \tilde{G}_0(r, r'; s)$$

$$= \frac{\gamma}{4\pi} [(-2\zeta^{-3} - 2\zeta^{-2}) e^{-\zeta} \boldsymbol{R}_0 \boldsymbol{R}_0$$

$$+ (\zeta^{-3} + \zeta^{-2} + \zeta^{-1}) e^{-\zeta} (\underset{\sim}{\boldsymbol{1}} - \boldsymbol{R}_0 \boldsymbol{R}_0)]$$

(3.18)

where

$$\boldsymbol{R} \equiv r - r', \qquad \boldsymbol{R}_0 \equiv \frac{r - r'}{|r - r'|}.$$

(3.19)

Consider now the MoM form of the response of an object of interest as in (3.13). For perfectly conducting objects in free space the elements of $(\tilde{\Gamma}_{n,m}(s))$ are simply related to the Green's functions above for the E-field integral equation, H-field integral equation, and other types of integral equations (see Chapt. 2). Note that these Green's functions are analytic functions of s except for $s=0$ (also, they do not exist for $R=0$). However, if the zones on the object are of finite size for a zoning or point-matching procedure then all the $\tilde{\Gamma}_{n,m}(s)$ must be entire functions except for finite order poles at $s=0$. Note that the elements of the forcing vector $(\tilde{I}_n(s))$ are entire functions corresponding to delta function excitation in time domain. If the object is not perfectly conducting then the medium parameters are assumed to be of sufficiently simple form so as to provide $(\tilde{\Gamma}_{n,m}(s))$ giving rise at most to finite order poles.

The solution of (3.13) is

$$
(\tilde{U}_n(s)) = \frac{1}{\det((\tilde{\Gamma}_{n,m}(s)))} (\tilde{d}_{n,m}(s)) \cdot (\tilde{I}_n(s))
$$
$$
\tilde{d}_{n,m}(s) = (-1)^{n+m} \det((\tilde{\Gamma}_{n,m}(s))')
$$
(3.20)

where $(\tilde{\Gamma}_{n,m}(s))'$ is the $(N-1)\times(N-1)$ matrix formed by deleting the n-th row and m-th column of $(\tilde{\Gamma}_{n,m}(s))$. The determinant is a sum of products of the matrix elements. Hence in the finite s plane the numerator matrices and $[\det((\tilde{\Gamma}_{n,m}(s)))]^{-1}$ have only pole singularities. Therefore $(\tilde{U}_n(s))$ has only pole singularities in the finite s plane. An object with such restrictions then has its response characterized by poles plus an entire function.

An important aspect of the MoM approach involves the convergence of the solution derived from a particular MoM representation. For example certain analytic thin-wire approximations do not yield more accurate results for arbitrarily large N because of certain inaccuracies in these approximations; there exists some optimum zone size in units of the wire radius [3.12]. Of course, one need not use such approximations. When MoM is employed to calculate the pole locations from (3.20) one must be assured of the accuracy (including choice of N) of the particular MoM form chosen.

There are various integral equations which have been used to determine the surface current density on perfectly conducting objects using MoM procedures. The E-field integral equation (EFIE) kernel is quite singular as $r \to r'$. Less singular is the H-field integral equation kernel as $r \to r'$. Both integral equations involve the Green's functions $\tilde{G}_0$ and $\nabla\tilde{G}_0$, respectively, with r and r' on the same surface. This difficulty is avoided by an extended-boundary-condition integral equation of electric or magnetic type for which the source (incident) fields are evaluated on a separate surface inside the body. In such cases the MoM matrix elements do not have to contend with $r \to r'$ so that a smoother kernel is numerically approximated. Whichever type of numerical MoM matrix is generated the general formulas for the pole factors (to be developed later) can be applied to it.

An essential feature of finite size objects in free space is that the $\tilde{\Gamma}_{n,\,m}(s)$ elements in the MoM form of solution all have $R = |r - r'|$ bounded by the maximum linear dimension of the object. This avoids infinitely large $\tilde{\Gamma}_{n,\,m}(s)$ in the left half plane, thus ensuring the analytic properties of the matrix elements. The finite size of the object is therefore essential to the occurrence of poles only in the finite s plane, whence finite-size objects in free space have special SEM properties. Avoiding the use of MoM one can use integral equation theory to investigate the analytic properties of the object response. This has been done for some finite-size objects. In particular, for perfectly conducting objects [3.3] and finitely conducting objects of finite size (with certain constraints on the constitutive parameters $(\sigma, \varepsilon, \mu)$ to guarantee sufficient conditions for only poles in the finite s plane) [3.7] it has been shown that the singularities associated with the object response are poles (of finite order) plus an entire function.

The outline of the proof due to MARIN [3.3] for the important case of a perfectly conducting body follows. Starting from the MFIE with kernel involving $\nabla \tilde{G}_0$ it can be shown that this kernel is a bounded operator analytic in the entire s plane. Next it is shown that the square of this kernel is a Hilbert-Schmidt operator. The MFIE is changed to one involving the squared operator. The integral equation is still of the second kind and Fredholm determinant theory is used. This gives a solution for the current which is an analytic function of s except for poles in the finite s plane. For this and other detailed derivations and proofs the reader should consult the cited references.

The perfectly-conducting finite-size object in free space approximates many scattering problems and is thus rather important. It has been shown [3.2] that a perfectly conducting sphere has only first order poles. It is conjectured that this is true of all finite-size perfectly-conducting objects and none of the many examples considered to date provides any exception to this conjecture. Resistive loaded bodies can have second-order poles [3.25] if the resistive loading parameters are carefully chosen.

3.2.4 Cases with More General Behavior in the s Plane

There are other types of objects and medium properties than those considered thus far. Such other problems can introduce new kinds of terms in the SEM representation.

First the analytical properties of the response function of an object situated between two perfectly conducting parallel planes have been considered [3.7]. Both poles and branch points now appear in the finite s plane. The branch point locations depend on the separation of the perfectly conducting parallel planes. A second type of problem concerns two-dimensional geometries, i.e., geometries independent of spatial translation in one direction, typically referred to as the z direction. The sources and fields, however, need not have this translation symmetry. In this case there are leaky modes (similar to natural modes) corresponding to poles in the p plane where

$$p^2 = \frac{s^2}{c^2} - \xi^2 \qquad\qquad (3.21)$$

and ξ is the Laplace transform variable corresponding to the longitudinal coordinate, the z axis. ξ can be interpreted as a propagation constant which for each pole in the p plane has a branch point and associated branch cut in the s plane. There is also a branch cut in the p plane along the negative real p axis. An example of this second type of problem is propagation on two parallel perfectly conducting uniform wires, as considered by MARIN [3.23]. Another example is that of a plane wave incident on an electrically thick perfectly conducting cylinder considered by SHAFER and KOUYOUMJIAN [3.22]. Depending on the form of excitation the leaky modes can appear in the solution in different ways, giving poles and/or branch cuts in the s plane. The analytic properties of the response functions for infinite objects can thus be more complicated than those of finite-size objects located in suitably restricted media. Detailed characteristics of the response will emerge from further studies. A third type of problem arises from s plane singularities associated with a physical medium through which a wave is propagating. Such a dispersive medium (perhaps including loss) can then introduce s plane singularities in the wave function, as discussed in FELSEN and MARCU-VITZ [3.50] (see also Sect. 1.6, and Chapt. 5).

3.3 SEM for Current and Charge Densities on Finite-Size Objects in Free Space

3.3.1 Some Definitions

Before proceeding to the derivations and formulas pertaining to the singularity expansion we provide definitions of some of the important SEM quantities.

s_α: natural frequency. This is a complex frequency, distinguished by the index α, for which the problem has a non-trivial solution without a forcing function. Each s_α depends only on the object parameters and not on the excitation or on spatial coordinates.

$v_\alpha(r)$:   natural mode. This is the non-trivial solution (vector, scalar, or dyadic) of the source-free (homogeneous) field equations at s_α which depends on position r and on the object parameters. Superscripts will indicate the appropriate electromagnetic quantity.

$\tilde{\eta}_\alpha(s)$:   coupling coefficient. This is the strength of the natural oscillation in terms of the object and incident wave parameters. It is independent of position.

$\mu_\alpha(r)$:   coupling vector. This is the non-trivial homogeneous solution (conjugate of the adjoint to v_α). When the field problem is phrased in the form of an integral equation, μ_α is associated with the transpose of the integral equation operator which is used in calculating the coupling coefficient. For symmetric integral equation operators (or matrices) μ_α is the same (within a scaling constant) as the natural mode v_α which solves the homogeneous equation on the righthand side.

f_p: incident waveform or source waveform with $p = 2, 3$ indicating polarization where required.

U_p: response to delta function excitation.

V_p: response to actual incident waveform ($\tilde{V}_p = \tilde{f}_p \tilde{U}_p$).

V_{po}:   object part of the response.

V_{pw}:   incident waveform (or other source) part of the response.

3.3.2 Form of Current Response in Complex Frequency Domain

The (Laplace transformed) surface currents induced by delta function excitation on finite-size perfectly conducting objects in free space are [3.2, 3, 9]

$$\tilde{U}_p^{(J)}(r,s) = \sum_\alpha \tilde{\eta}_\alpha(\mathbf{1}_0,s) v_\alpha^{(J)}(r)(s - s_\alpha)^{-m_\alpha} + \tilde{W}_p^{(J)}(\mathbf{1}_0,r,s) \qquad (3.22)$$

with m_α being a positive integer, where the equations for the pole factors are given in the next subsection and where $\tilde{W}_p$ is an entire function which will normally be suppressed for purposes of the present discussion. It is recalled that superscripts identify the observable quantity (field, charge, current) under consideration. The notation $\tilde{\eta}_\alpha(\mathbf{1}_0,s)$ indicates that $\tilde{\eta}_\alpha$ depends on the incident wave direction as well as on

s. According to the Mittag-Leffler theorem [3.46, 3], each pole term in the infinite series requires an entire function to guarantee convergence of the series; these additional entire functions have not been shown explicitly. However, the coupling coefficient $\tilde{\eta}_\alpha$, if regarded as a function s, does allow inclusion of an entire function with each term. As will be seen later there is a certain ambiguity inherent in the coupling coefficient except at $s = s_\alpha$. This ambiguity is in part relatable to the Mittag-Leffler representation. The form of the pole factorization in (3.22) is derived in Subsection 3.3.3.

The current density for plane wave incidence is constructed as

$$
\begin{aligned}
\tilde{J}(r,s) &= \tilde{J}_2(r,s) + \tilde{J}_3(r,s) \\
\tilde{J}_p(r,s) &= E_0\, \psi\, \tilde{f}_p(s)\, \tilde{U}_p^{(J)}(r,s)\,, \qquad p = 2,3
\end{aligned}
\tag{3.23}
$$

where ψ is a normalizing constant and E_0 is some convenient source electric field amplitude. Where needed 1_0 is the direction of incident wave propagation and 2_0 and 3_0 are polarizations orthogonal to each other and to 1_0.

One important feature of the singularity expansion is the separation of object and waveform $(\tilde{f}_p(s))$ singularities when these are distinct. The normalized response to the incident waveform is

$$
\begin{aligned}
\tilde{V}_p^{(J)}(r,s) &= \tilde{f}_p(s)\, \tilde{U}_p^{(J)}(r,s) \\
&= \tilde{V}_{p_0}^{(J)}(r,s) + \tilde{V}_{p_w}^{(J)}(r,s)
\end{aligned}
\tag{3.24}
$$

Restricting consideration to the case of first-order poles due to the object and neglecting $\tilde{W}_p$, the object and waveform parts are, respectively,

$$
\begin{aligned}
\tilde{V}_{p_0}^{(J)}(r,s) &= \sum_\alpha \tilde{f}_p(s_\alpha)\, \tilde{\eta}_\alpha(1_0,s)\, v_\alpha^{(J)}(r)(s - s_\alpha)^{-1} \\
\tilde{V}_{p_w}^{(J)}(r,s) &= \sum_\alpha \tilde{\eta}_\alpha(1_0,s)\, v_\alpha^{(J)}(r)\, \frac{\tilde{f}_p(s) - \tilde{f}_p(s_\alpha)}{s - s_\alpha} \\
&= \sum [\text{singularities of } \tilde{f}_p(s)]\, \tilde{U}_p^{(J)}(r,s)\, \Big|_{\substack{s \text{ at } f_p \\ \text{singularities}}} \\
&\quad + \text{possible entire function,}
\end{aligned}
\tag{3.25}
$$

where the last form is found as a singularity expansion over the singularities of the incident waveform and where the object and waveform singularities have been assumed to be separated in the s plane. By this separation one exhibits the natural frequencies and modes of the object which are independent of the incident wave parameters. Only the strengths of the natural oscillations are affected by the incident waveform, and then in the form of the simple factor $\tilde{f}_p(s_\alpha)$.

3.3.3 Formulas and Derivations for Terms Associated with Poles

Note the factorization associated with pole terms in (3.22), one of the most powerful features of SEM for understanding responses and simplifying parameter studies. This aspect is now considered in some detail.

Starting with the integral equation (3.12) for the delta function response

$$\langle \tilde{\underline{\Gamma}}(r,r';s); \tilde{U}(r',s)\rangle = \tilde{I}(r,s) \tag{3.26}$$

first find the natural frequencies and natural modes from

$$\langle \tilde{\underline{\Gamma}}(r,r';s_\alpha); v_\alpha(r')\rangle = 0. \tag{3.27}$$

In MoM form this equation is

$$(\tilde{\Gamma}_{n,m}(s_\alpha)) \cdot (v_n)_\alpha = (0_n) \tag{3.28}$$

which gives an equation for the natural frequencies as

$$\det((\tilde{\Gamma}_{n,m}(s_\alpha))) = 0 \tag{3.29}$$

(see Sect. 3.6 for a discussion of numerical procedures for solving this equation). Having found the natural frequencies and natural modes one can find the coupling vectors from

$$\langle \mu_\alpha(r); \tilde{\Gamma}(r,r';s_\alpha)\rangle = 0 \tag{3.30}$$

or

$$(\mu_n)_\alpha \cdot (\tilde{\Gamma}_{n,m}(s_\alpha)) = (0_n). \tag{3.31}$$

Now expand the kernel of the integral equation in a power series around $s = s_\alpha$ as

$$\tilde{\underline{\Gamma}}(r,r';s) = \sum_{l=0}^{\infty} (s-s_\alpha)^l \Gamma_{l_\alpha}(r,r')$$

$$\tilde{\underline{\Gamma}}_{l_\alpha}(r,r') = \frac{1}{l!} \frac{\partial^l}{\partial s^l} \tilde{\underline{\Gamma}}(r,r';s)\Big|_{s=s_\alpha} \tag{3.32}$$

Similarly expand

$$\tilde{I}(r,s) = \sum_{l=0}^{\infty} (s-s_\alpha)^l I_{l_\alpha}$$
$$I_{l_\alpha} = \frac{1}{l!} \frac{\partial^l}{\partial s^l} I(r,s)\Big|_{s=s_\alpha} \tag{3.33}$$

Write the response by separating the pole at s_α (assuming a first-order pole) as

$$\tilde{U}(r,s) = \tilde{\eta}_\alpha(1_0,s_\alpha)v_\alpha(r)(s-s_\alpha)^{-1} + \tilde{U}'(r,s) \tag{3.34}$$

where $\tilde{U}'(r,s)$ is analytic in the neighborhood of s_α. The vector coefficient of the pole is written in a factored form which is justified below.

Substitute these expansions around $s=s_\alpha$ into the integral equation and group the results according to powers of $s-s_\alpha$. The coefficient of the $(s-s_\alpha)^{-1}$ term gives at s_α

$$\langle \underset{\sim}{\Gamma}_{0_\alpha}(r,r'); \tilde{\eta}_\alpha(1_0,s_\alpha)v_\alpha(r) \rangle = 0, \tag{3.35}$$

which is consistent with (3.27). The coefficient of the $(s-s_\alpha)^0$ term gives at s_α

$$\begin{aligned} &\langle \underset{\sim}{\Gamma}_{0_\alpha}(r,r'); \tilde{U}'(r',s_\alpha) \rangle \\ &+ \langle \underset{\sim}{\Gamma}_{1_\alpha}(r,r'); \tilde{\eta}_\alpha(1_0,s_\alpha)v_\alpha(r') \rangle = I_{0_\alpha}(r). \end{aligned} \tag{3.36}$$

Operating on the left by $\mu_\alpha(r)$ and noting that the first term in (3.36) is thereby made to be zero we have

$$\langle \mu_\alpha(r); \underset{\sim}{\Gamma}_{1_\alpha}(r,r'); \tilde{\eta}_\alpha(1_0,s_\alpha)v_\alpha(r') \rangle = \langle \mu_\alpha(r); I_{0_\alpha}(r) \rangle. \tag{3.37}$$

This equation is solved for the coupling coefficient as

$$\tilde{\eta}_\alpha(s_\alpha) = \frac{\langle \mu_\alpha(r); I_{0_\alpha}(r) \rangle}{\langle \mu_\alpha(r); \underset{\sim}{\Gamma}_{1_\alpha}(r,r'); v_\alpha(r) \rangle} = \text{coupling coefficient at } s_\alpha. \tag{3.38}$$

The essence of this derivations is the expansion near s_α, substituting the series in the integral equation, collecting terms by powers of $s-s_\alpha$, and judicious left operation by μ_α. Note that the natural modes and coupling vectors are unnormalized but that this has no effect on the product of the coupling coefficient and natural mode.

3.3.4 Forms of Coupling Coefficients

Equation (3.38) gives the coupling coefficient at $s=s_\alpha$; this provides the correct residue at the α-th pole. For other values of s the coupling coefficient can have various forms chosen for convenience and/or for desirable rate of convergence of the sum. Note that any entire function with a zero at $s=s_\alpha$ can be added to $\tilde{\eta}_\alpha(s_\alpha)$ to give an $\tilde{\eta}_\alpha(s)$ satisfying (3.26). Hence the choice of a particular form of $\tilde{\eta}_\alpha(s)$ is related to the presence or lack of a remaining entire function ($\tilde{W}_p$ in (3.22)).

Since an entire function arises from the Laplace transform and since it has no singularities in the finite s plane it must rise and fall in the time domain faster than an exponential. Such a time function is then limited in importance to early times since the damped sinusoids will dominate at late times. The late-time forms of the pole terms are then somewhat insensitive to the form of $\tilde{\eta}_\alpha$. In the frequency domain, however, the choice of $\tilde{\eta}_\alpha$ can affect the pole terms across the spectrum.

There are two classes of coupling coefficients which have been found useful. The first and simplest of these, referred to as "class 1", for turn-on at t', is given by [3.2, 9]

$$\tilde{\eta}_\alpha(s) = e^{(s_\alpha - s)t'} \frac{\langle \mu_\alpha(r); I_{0\alpha}(r) \rangle}{\langle \mu_\alpha(r); \Gamma_{1\alpha}(r, r'); v_\alpha(r) \rangle}. \qquad (3.39)$$

Here t' is a "turn-on time" chosen for convenience and convergence. It might be the time at which the incident or other source field is first applied anywhere on the object. It has been shown [3.2] that such a choice gives the solution for the current density induced on a sphere by an incident delta-function plane wave without the necessity of an additional entire function. Causality considerations indicate that if t' is chosen after the first time at which a response at some particular r on the object is known to occur then an additional entire function term is certainly required.

"Class 2", or convolution form coupling coefficients, are given by [3.2, 3, 9]

$$\tilde{\eta}_\alpha(s) = \frac{\langle \mu_\alpha(r); \tilde{I}(r, s) \rangle}{\langle \mu_\alpha(r); \Gamma_{1\alpha}(r, r'); v_\alpha(r') \rangle}. \qquad (3.40)$$

The s dependence of $\tilde{\eta}_\alpha$ here comes from $\tilde{I}(s)$. In the time domain this corresponds to smoothing out the rise time of the α-th pole term by a convolution.

Class 2 coupling coefficients are more complicated to calculate than class 1 coupling coefficients. However, class 2 coupling coefficients give smoother early-time results for a finite number of poles when these are included in a numerical summation scheme because of the smoother rise of the resulting pole terms in time domain. For plotting the coupling coefficient as a function of the object parameters and incident wave parameters it is best to just use $\tilde{\eta}_\alpha(s_\alpha)$ which is unique. Note that both classes give the same pole residues since $\tilde{I}(s_\alpha) = I_{0_\alpha}$.

3.3.5 Form of Current Response in Time Domain

The formulas in Subsection 3.3.2 can be written in time domain form. Neglecting a possible additional entire function the class 1 coupling coefficients give time domain forms as

$$U_p^{(J)}(r,t) = u(t-t') \sum_\alpha \tilde{\eta}_\alpha(1_0, s_\alpha) v_\alpha^{(J)}(r) e^{s_\alpha t}$$

$$V_{p0}^{(J)}(r,t) = u(t-t') \sum_\alpha \tilde{f}_p(s_\alpha) \tilde{\eta}_\alpha(1_0, s_\alpha) v_\alpha^{(J)}(r) e^{s_\alpha t}, \tag{3.41}$$

where the occurrence of only first-order poles has been assumed and where $u(t-t')$ is the Heaviside unit step function. The waveform part of the response depends on the form chosen for $f_p(t)$, the incident waveform or other forcing function. For certain excitations the waveform part of the response is rather simple. One interesting example is provided by step function excitation for which $\tilde{f}_p(s) = s^{-1}$, whence from (3.24) and (3.2)

$$V_{p_w}^{(J)}(r,t) = u(t-t') \tilde{U}_p^{(J)}(r,0) \tag{3.42}$$

where the $\tilde{U}_p$ term is the static response which can be calculated by static techniques (Laplace equation). This comes from the waveform singularity at $s=0$, from $\exp(s_\alpha t')/s$ (step function excitation at time t') consistent with class 1 coupling coefficients.

The class 2 coupling coefficients give the time domain response (neglecting $\tilde{W}_p$ and assuming the occurrence of first-order pole only) as

$$U_p^{(J)}(r,t) = \sum_\alpha [\langle \mu_\alpha(r); \Gamma_{1_\alpha}(r,r'); v_\alpha(r') \rangle]^{-1} [\langle \mu_\alpha(r); I(r,t) \rangle]$$

$$* [u(t) e^{s_\alpha t}] v_\alpha^{(J)}(r)$$

$$V_{p0}^{(J)}(r,t) = \sum_\alpha \tilde{f}_p(s_\alpha) [\langle \mu_\alpha(r); \Gamma_{1_\alpha}(r,r'); v_\alpha(r') \rangle]^{-1} \tag{3.43}$$

$$[\langle \mu_\alpha(r); I(r,t) \rangle] * [u(t) e^{s_\alpha t}] v_\alpha^{(J)}(r),$$

where an asterisk $*$ between two time-domain terms indicates convolution. The individual terms are seen to be more complicated than for class 1, but the early-time response is rounded off. The waveform part of the response is likewise more complicated.

Note that for the real-valued time-domain response one can order the poles into three groups:

let

$$
\begin{aligned}
\alpha &\equiv \alpha_+ \quad \text{for} \quad \text{Im}\{s_\alpha\} > 0 \\
\alpha &\equiv \alpha_0 \quad \text{for} \quad \text{Im}\{s_\alpha\} = 0 \\
\alpha &\equiv \alpha_- \quad \text{for} \quad \text{Im}\{s_\alpha\} < 0 .
\end{aligned}
\tag{3.44}
$$

Then the above summations in the time domain can be phrased so as to include twice the real part for the α_+ terms, retain the α_0 terms, and omit the α_- terms.

3.3.6 Charge Density

An equally important part of the object response is the charge density, particularly for objects with the current density concentrated on a surface. The charge natural modes are found from [3.26]

$$
v_\alpha^{(\varrho)}(r) = -\frac{a_\alpha}{\gamma_\alpha} \nabla \cdot v_\alpha^{(J)}(r), \qquad \gamma_\alpha \equiv \frac{s_\alpha}{c}
\tag{3.45}
$$

where a_α is a convenient length scaling factor. The charge response in the complex frequency domain (for first-order poles and neglecting any additional entire function) is found by applying (3.45) to the earlier results for the current

$$
\begin{aligned}
\tilde{U}_p^{(\varrho)}(r,s) &= \sum_\alpha \tilde{\eta}_\alpha(1_0,s) v_\alpha^{(\varrho)}(r) \frac{s_\alpha}{s a_\alpha}(s-s_\alpha)^{-1} \\
\tilde{\varrho}_p(r,s) &= E_0 \frac{\xi}{c} \tilde{f}_p(s) \tilde{U}_q^{(\varrho)}(r,s) \\
\tilde{V}_p^{(\varrho)}(r,s) &= \tilde{f}_p(s) \tilde{U}_p^{(\varrho)}(r,s) = \tilde{V}_{p_0}^{(\varrho)}(r,s) + \tilde{V}_{p_w}^{(\varrho)}(r,s) \\
\tilde{V}_{p_0}^{(\varrho)}(r,s) &= \sum \tilde{f}_p(s_\alpha) \tilde{\eta}_\alpha(1_0,s) v_\alpha^{(\varrho)}(r) a_\alpha^{-1}(s-s_\alpha)^{-1} \\
\tilde{V}_{p_w}^{(\varrho)}(r,s) &= \sum_\alpha \tilde{\eta}_\alpha(1_0,s) v_\alpha^{(\varrho)}(r) a_\alpha^{-1} \frac{\tilde{f}_p(s) - \tilde{f}_p(s_\alpha)}{s-s_\alpha} .
\end{aligned}
\tag{3.46}
$$

In the time domain, class 1 coupling coefficients give the object response as

$$V_{p0}^{(\varrho)}(r,t) = u(t-t') \sum_{\alpha} \tilde{f}_p(s_\alpha) \tilde{\eta}_\alpha(1_0, s_\alpha) v_\alpha^{(\varrho)}(r) a_\alpha^{-1} e^{s_\alpha t} . \tag{3.47}$$

For step excitation the waveform part of the response is

$$V_{pw}^{(\varrho)}(r,t) = u(t-t') \tilde{U}_p^{(\varrho)}(r,0) \tag{3.48}$$

where the $\tilde{U}_p$ term is the electrostatic response. The class 2 object response is

$$V_{p0}^{(\varrho)}(r,t) = \sum_{\alpha} \tilde{f}_p(s_\alpha) [\langle \mu_\alpha(r); \Gamma_{1_\alpha}(r,r'); v_\alpha(r') \rangle]^{-1_{a_\alpha}-1}$$
$$[\langle \mu_\alpha(r); I(r,t) \rangle] * [u(t) e^{s_\alpha t}] v_\alpha^{(\varrho)}(r) . \tag{3.49}$$

3.3.7 Numerical Examples

Induced current and charge densities on various types of objects have been calculated using SEM procedures. These specific examples have served to develop and illustrate some of the important features of the SEM form of the response and to establish its practical applicability. Since certain theoretical questions involving the additional entire function are not fully resolved these examples give evidence as to whether or not such a term is needed in special types of problems.

The first case treated was the perfectly conducting sphere as in Fig. 3.3. This case could be solved analytically. It was shown [3.2] that the poles were all of first order and that class 1 coupling coefficients with $t' = -a/c$, where a is the sphere radius, could be used without an additional entire function, for the responses to an incident plane wave with either delta-function or step-function time dependence. MAR-TINEZ et al. [3.6] subsequently performed numerical calculations of the SEM quantities and the resulting waveforms. The lowest-order natural frequencies are shown in Fig. 3.4 with identification as to which natural frequencies belong to natural modes with zero surface charge density. Figures 3.5 and 3.6 show the step function response convergence for various "arcs" of natural frequencies (sets of natural frequencies associated with each term in the eigenfunction solution), with the pole at $s=0$ included; θ' is selected as $3\pi/4$ for these figures. One observes very rapid convergence in the time domain for intermediate times (t of the order of a/c) and late times ($t \gg a/c$) with not very many (say

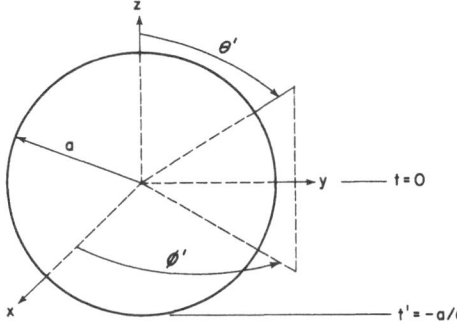

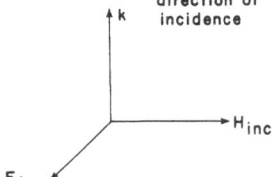

Fig. 3.3. Plane wave incident on perfectly conducting sphere. The angle coordinates (θ', ϕ') locate points on the surface

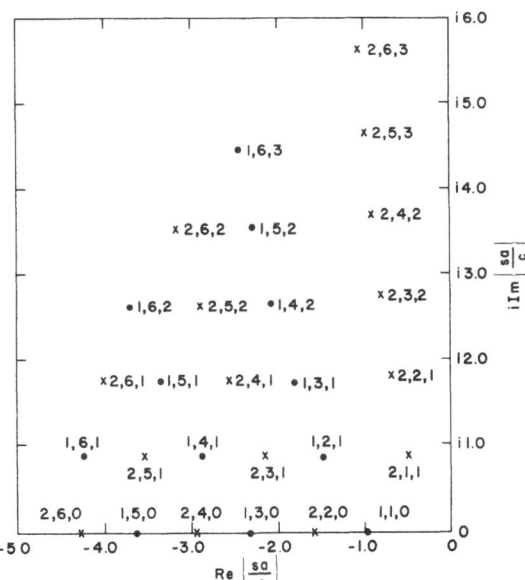

Fig. 3.4. Natural frequencies of the perfectly conducting sphere in the complex plane (sa/c). For use with exterior incident wave, $1 \le n \le 6$, index set is three numbers q, n, n'. $q = 1$ denotes the case where there is no surface charge density. $q = 2$ denotes the case of non-vanishing charge density. The index n identifies the pole grouping associated with an n-th order Bessel function; the n' identifies a particular pole within that group

10) pole pairs. In the frequency domain the number of pole pairs required for convergence of the sum of poles to the exact solution was proportional to $(ka)^2$. However, the summation order was taken to include poles in a portion of the s plane with approximately constant ratio of $\text{Re}\{s\}$ to $\text{Im}\{s\}$ extent from the origin. Beside the pole at $s = 0$ all poles associated with the Bessel functions up to order N are included. (See [Ref. 3.41, Sect. 9.23] for scattering from a perfectly conducting sphere. The index on the Bessel functions is n so all poles for $1 \leq n \leq N$ plus the pole at zero is included.) Checks were made by including single-pole pairs from the next arcs (order $N + 1$) with still good convergence. The index n' is used to specify the zeros for a given index pair q, n.

The first numerical study using MoM techniques ([3.47] and Sect. 3.5) was that of finite-length thin-wire cylinder by TESCHE [3.5] with a geometry as indicated in Fig. 3.7. For a rather thin cylinder ($d/L = 0.01$) the pole positions are as indicated in Fig. 3.8. As the wire is made thicker the poles near the $i\text{Im}\{s\}$ axis move farther into the left half plane. Figure 3.9 shows the first several natural modes (smallest $|s_\alpha|$ in the first layer near the $i\text{Im}\{s\}$ axis) and Fig. 3.10 exhibits the variation of the coupling coefficients as a function of the angle of incidence with the $t = 0$ time reference at the wire center. Figure 3.11 displays the convergence of the step response using the class 2 form of the coupling coefficients, as a function of the number of pole pairs (taken

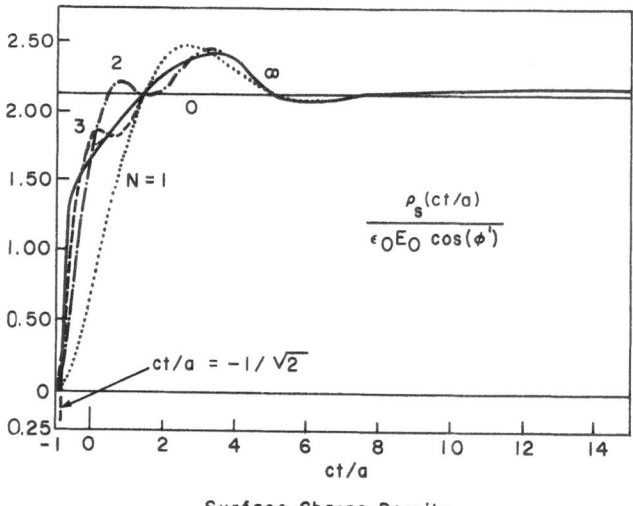

Surface Charge Density

Fig. 3.5. Step function response for surface current and charge densities on a perfectly conducting sphere, with $\theta' = 3\pi/4$

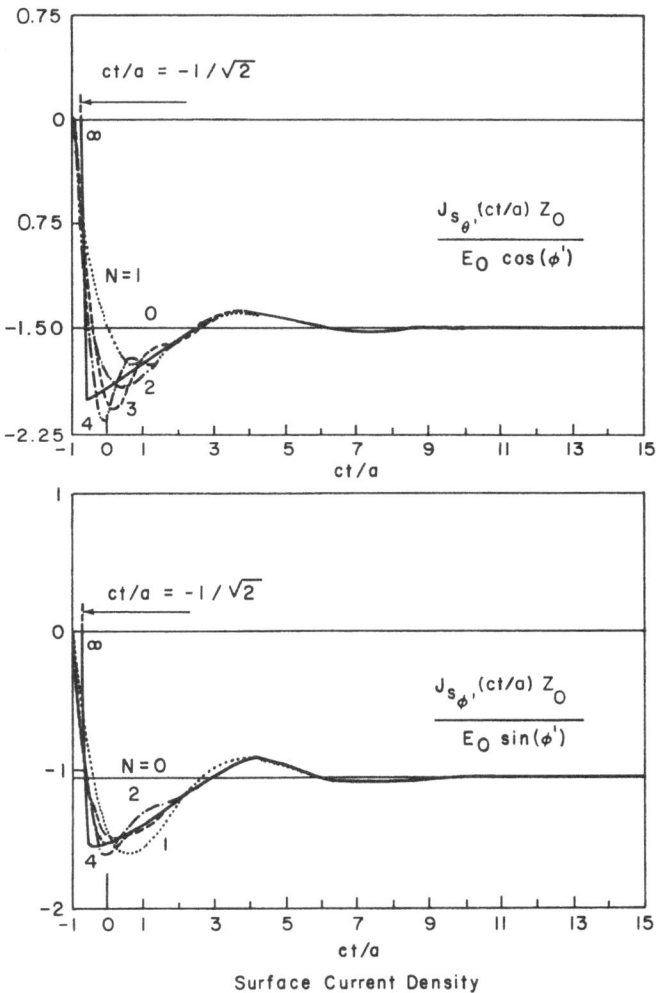

Fig. 3.5 (continued)

only from the layer near the $i\text{Im}\{s\}$ axis). For about 10 pole pairs the results are virtually indistinguishable from the results obtained by inverting the matrix in (3.13) at a number of frequencies sufficiently large to observe numerical convergence and performing a numerical inverse Fourier transform. As expected the most noticeable differences are at early times where more terms are needed. At late times the first few pole pairs are all that is required for an accurate and simple representation. These thin-wire results have been used by BARNES et al. [3.11, 14] to efficiently characterize the transient response of cylindrical dipole receiving antennas.

A thin prolate spheroid is similar to a thin wire. Marin [3.4] has studied numerically a prolate spheroid as in Fig. 3.12, using its property as a rotationally symmetric body. The axially symmetric modes were considered. This object allowed comparison between the results for a thin body and the analytical results for a sphere. As shown in Fig. 3.13 the natural frequencies vary continuously from the case of the sphere to that of a thin prolate spheroid where the results compare favorably to those of Tesche [3.5]. About 10 pole pairs were required here for good results for the step function response for a thin prolate spheroid.

Another rotationally symmetric object being considered is a circular aperture in an infinite plane, or equivalently the complementary problem of a perfectly conducting circular disk. Taylor et al. [3.10] have formulated a numerical type of solution.

The results for isolated thin objects have been extended to two thin perfectly conducting cylinders separated by a symmetry plane. This case includes the configuration of one such cylinder in proximity to a perfectly conducting infinite plane. Marin [3.24] has considered two identical colinear perfectly conducting cylinders, as illustrated in Fig. 3.14. This type of problem illustrates natural-frequency splitting

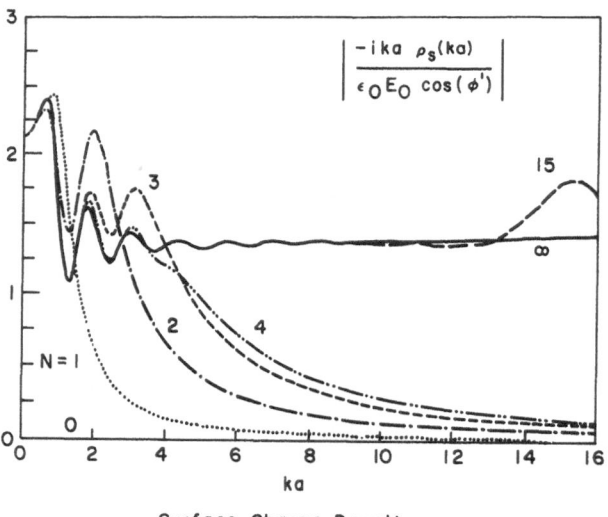

Surface Charge Density

Fig. 3.6. Step function response times -ika in the frequency domain for surface current and charge densities on a perfectly conducting sphere, with $\theta' = 3\pi/4$

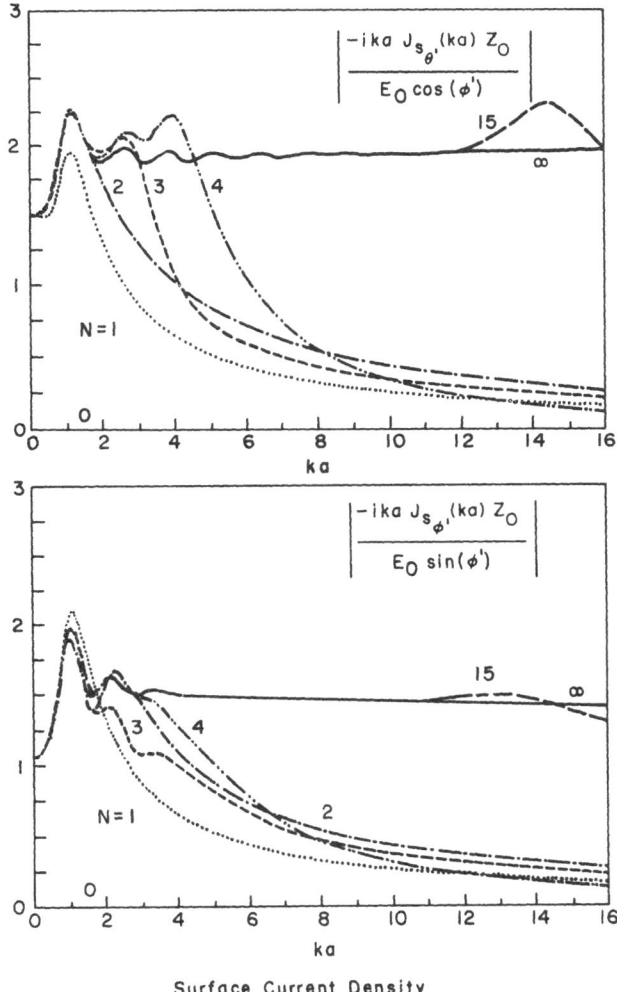

Surface Current Density

Fig. 3.6 (continued)

as the wire separation is varied, a phenomenon similar to energy-level splitting in quantum mechanics as two atoms (or other systems) are brought together. The splitting for the lowest-order pole is shown in Fig. 3.15, with one s_α associated with each of the two natural modes. Since there is a plane of symmetry there are two separate parts to the solution, symmetric and antisymmetric, and these symmetries are present in the natural modes [3.1]. The lowest-order natural modes are plotted in Fig. 3.16 and 3.17. Note that U is the normalized position

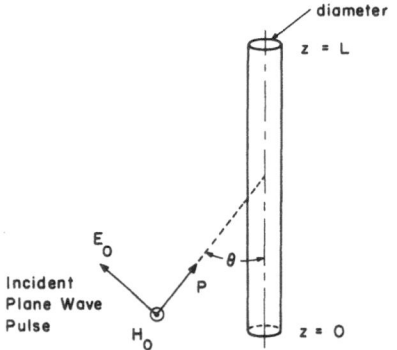

Fig. 3.7. Geometry of the wire scatterer and incident field

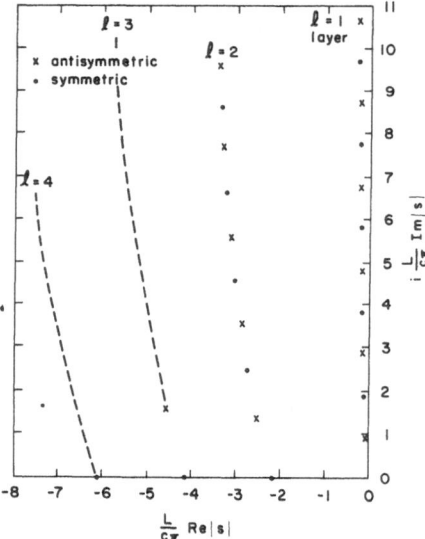

Fig. 3.8. Pole locations in the complex frequency plane for the thin wire of $d/L=0.01$. The poles are grouped into layers denoted by l

along the wire for $z>0$; $U=1$ corresponds to $z=d+2h$ and $U=-1$ corresponds to $z=d$. The study of this problem was continued by SHUM-PERT [3.27] including effects of cylinder inclination. One interesting effect he observed is shown in Fig. 3.18. As the cylinder spacing from the ground plane is changed there is an interchange of poles near

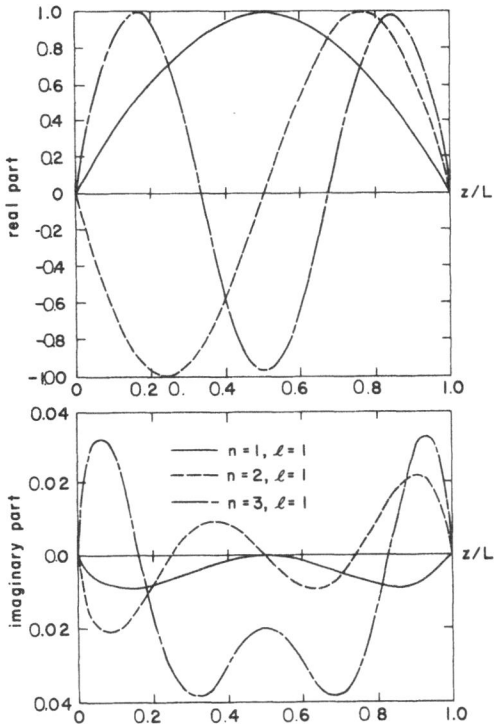

Fig. 3.9. Plots of the real and imaginary parts of the first three normalized natural modes for the thin wire, $d/L = 0.01$. $l = 1$ denotes the first layer (near the $i\ \mathrm{Im}\{s\}$ axis). n denotes which pole in this layer

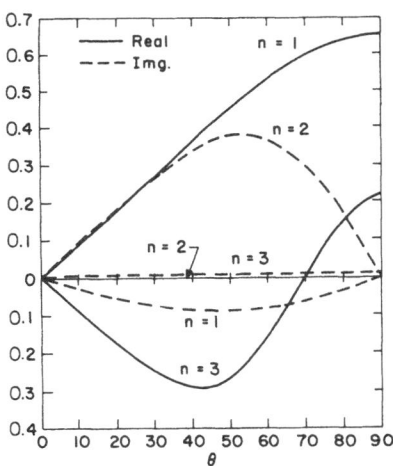

Fig. 3.10. Plots of the coupling co-efficients as a function of the angle of incidence θ for the first three poles in the $l = 1$ layer for the thin wire, $d/L = 0.01$

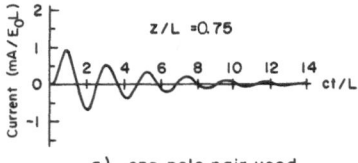

a) one pole pair used

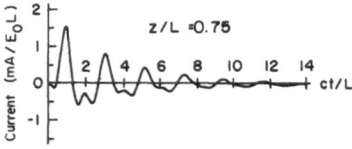

b) two pole pairs used

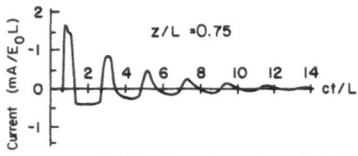

c) Fourier Inversion Solution

Fig. 3.11 a–c. The unit step response at $z/L=0.75$ on the thin wire for one or two pole pairs compared to the Fourier inversion solution. The angle of incidence is $\theta=30°$, and $d/L=0.01$. Only the $l=1$ poles are considered

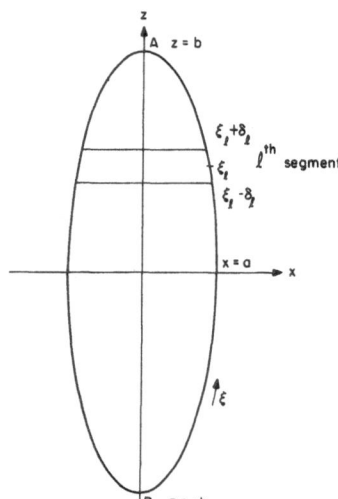

Fig. 3.12. A prolate spheroid with the segments and sample points used in the numerical calculations

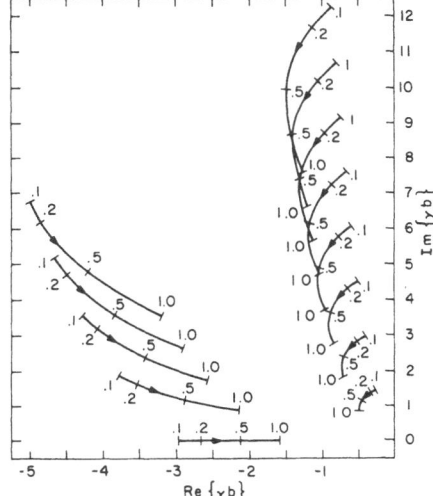

Fig. 3.13. The locus of natural frequencies for the prolate spheroid when $0.1 \le a/b \le 1$. The location of the natural frequencies for $a/b = 0.1, 0.2, 0.5, 1$ is indicated on the curves. The arrows indicate the direction in which the natural frequencies move for increasing values of a/b. Here $\gamma = s/c$

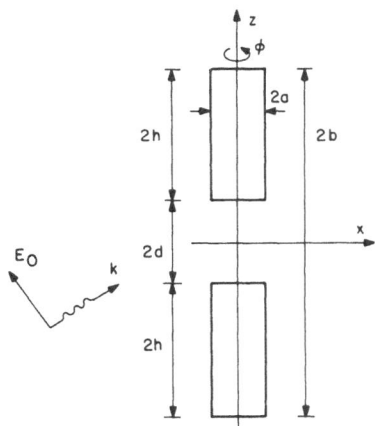

Fig. 3.14. Electromagnetic interaction of two colinear cylinders and a plane wave

the position of the free space pole with a new pole moving in to replace the old one. SHUMPERT's results are for the case of a ground plane, giving antisymmetric type modes. These cited results of MARIN and SHUMPERT show some of the questions which can be posed in an

160

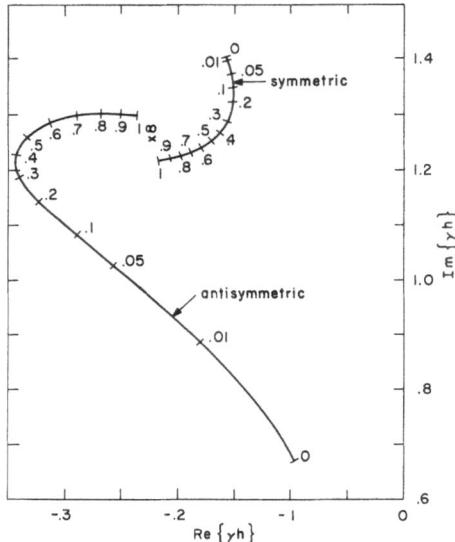

Fig. 3.15. Loci of the first natural frequency for two colinear cylinders, for different values of $0 < d/h < 1$

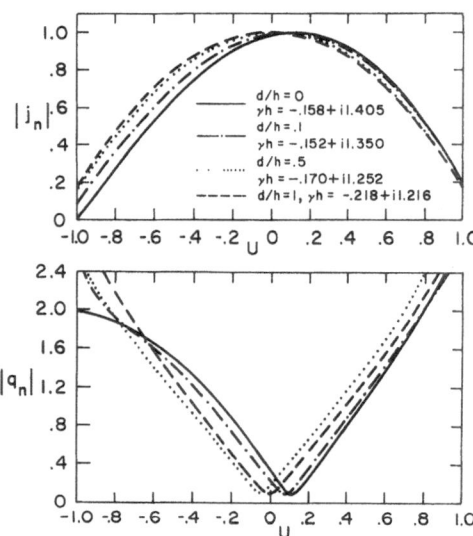

Fig. 3.16. Current and charge distributions of the first symmetric natural mode on two colinear cylinders, for $a/h = 0.1$ and different values of d/h

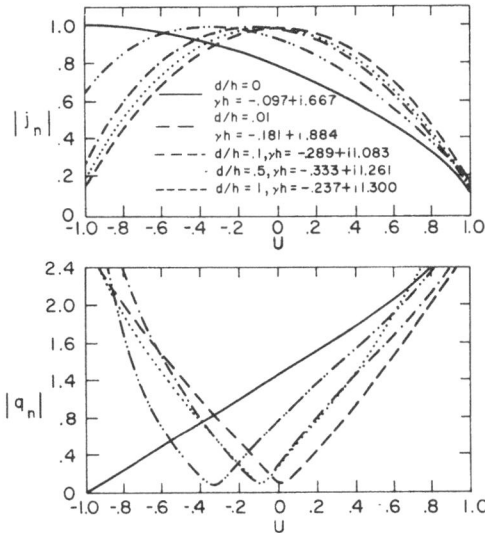

Fig. 3.17. Current and charge distributions of the first antisymmetric natural mode on two colinear cylinders, for $a/h = 0.1$ and different values of d/h

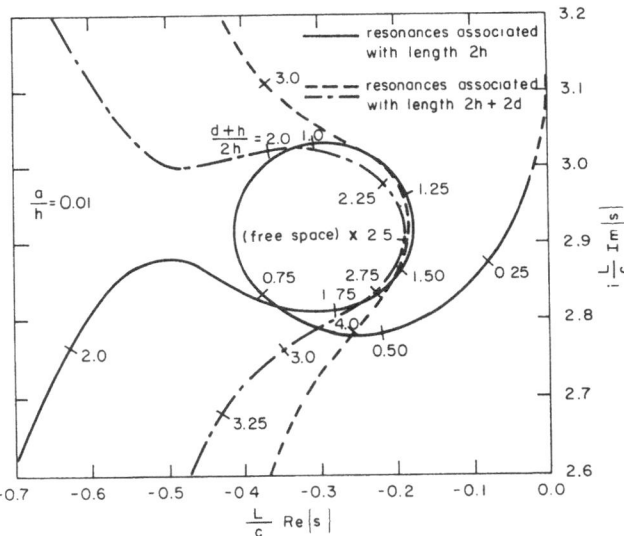

Fig. 3.18. Trajectory of the antisymmetric singularities associated with the first resonance of the scatterer without ground plane (SHUMPERT)

SEM context, suggesting new physical insights into the behavior of the electromagnetic response.

More complicated wire structures can be considered. WILTON and UMASHANKAR [3.13, 16] have treated a thin bent-wire structure using MoM. Again good convergence was observed. One interesting feature of their work was a comparison to direct matrix inversion in both frequency and time domains. Figure 3.19 shows a frequency domain comparison obtained by direct matrix inversion at each frequency; note the good agreement with about 10 poles. CROW et al. [3.15] considered a crossed thin cylinder configuration. This is a crude model of an aircraft, but it is a good starting point to illustrate some of the complicated electromagnetic response. Figure 3.20 illustrates the first few symmetric (fuselage) natural modes for the current, where current is positive toward l'_1 and $+l_2$ (for the wing). Note the different location of the maximum fuselage current for the first two modes. Figure 3.21 shows the first few antisymmetric natural modes for the current; these are on the wing. All these studies indicate good convergence for step excitation with five to ten pole-pair terms.

Class 1 coupling coefficients have been used in some of the examples discussed above and class 2 in others, both with success in the sense of avoiding an additional entire function for the cases considered. The number of terms required for good intermediate and late time response under step function excitation was about 10 or even fewer. For the step response just the first pole pair makes a rough but useful approximation.

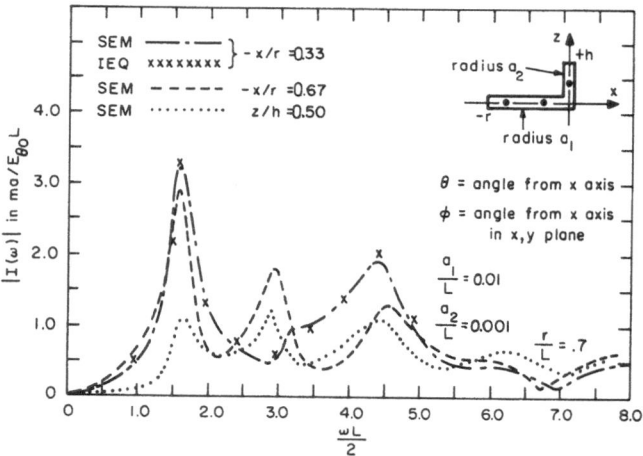

Fig. 3.19. Variation of current on the bent wire as a function of frequency for a plane wave incident, with $\theta = 45°$, $\phi = 45°$, E_θ-polarization there, $L = r + h$

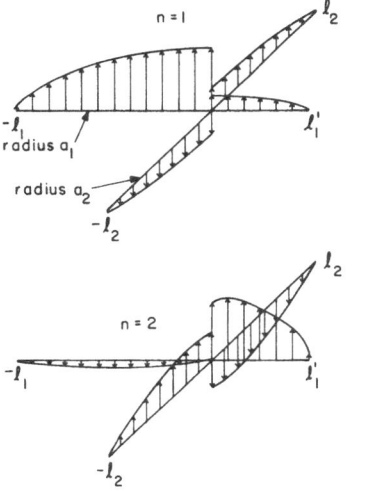

$a_1 = a_2$; $2l_2/(l_1 + l_1') = 1.0$; $l_1'/l_1 = 0.5$; $2l_2/a_2 = 20.0$

Fig. 3.20. Real part of the symmetric current modes on thin crossed wires

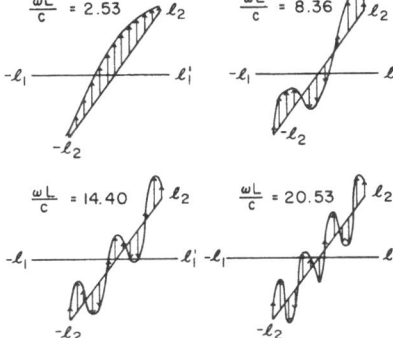

Fig. 3.21. Real part of the antisymmetric current modes on thin crossed wires

Most of the results to date for specific structure shapes have been obtained using numerical techniques of the general form of MoM. While such a general numerical approach is important for handling complicated structures it may create difficulties with numerical accuracy and also involves limitations on physical insight. Where tractable, analytical approximations are useful to increase insight and improve accuracy, thus serving as a check on, and a stimulus for improvement of, more general numerical methods. For thinwire structures the SEM quantities can be approximated analytically. LEE and LEUNG [3.8] have obtained approximations for the first few natural frequencies of a finite-length thin perfectly conducting cylinder. LEE and MARIN [3.19, 28] have recently developed analytical approximations for natural frequencies, natural modes, coupling coefficients, and other terms (such as static terms) in the SEM representation of various thin-wire structures including cylinders of varying radius and bent wire structures.

3.4 SEM for Fields Radiated or Scattered from Finite-Size Bodies in Free Space

3.4.1 Natural Modes

SEM applies to fields (potentials, etc.) as well as current and charge densities on objects. For finite size objects, the fields have properties similar to that associated with complex natural frequencies, natural modes, and possible entire functions for the currents and charge densities. One can construct an appropriate set of natural modes for the field quantities [3.26] as

$$
\begin{aligned}
v_\alpha^{(E)} &= -c_\alpha \gamma_\alpha^2 \langle \underset{\sim}{G}_{0_\alpha} ; v_\alpha^{(J)} \rangle \\
v_\alpha^{(H)} &= c_\alpha \gamma_\alpha \langle \nabla G_{0_\alpha} \times v_\alpha^{(J)} \rangle \\
v_\alpha^{(F_q)} &= -c_\alpha \gamma_\alpha^2 \langle \underset{\sim}{G}_{0_\alpha} ; v_\alpha^{(K_q)} \rangle + q\, i\, c_\alpha \gamma_\alpha \langle \nabla \tilde{G}_{0_\alpha} \times v_\alpha^{(K_q)} \rangle
\end{aligned}
\tag{3.50}
$$

where c_α is a convenient normalizing constant and the Green's functions are discussed in Subsection 3.2.3. The integration is over the volume occupied by the current density [see (3.11), etc., for definition of notation].

One difficulty with the use of natural modes for the fields is that they grow exponentially as $r \to \infty$, due to Re$\{s_\alpha\} < 0$, with $-s_\alpha r/c$ in the exponent of the Green's functions. This leads to the consideration of retarded and far field natural modes to circumvent this difficulty for large r. For r near the object, the simple natural modes as in equations (3.50) are appropriate.

For F_q or E the natural modes are the same as the current density natural modes in Section 3.3 if an additional coefficient of $Z_0/(c_\alpha \gamma_\alpha)$ is included. For H the additional coefficient is $1/(c_\alpha \gamma_\alpha)$. The expansions then become field expansions instead of current density expansions.

3.4.2 Retarded Natural Modes

Retarded natural modes are defined by [3.26]

$$v_{ret_\alpha}^{(X)}(r) \equiv e^{\gamma_\alpha r} v_\alpha^{(X)}(r), \qquad \gamma_\alpha = s_\alpha/c, \tag{3.51}$$

where X denotes the electromagnetic quantity of interest. This type of natural mode decays as $r \to \infty$. It corresponds to a shift from time t to retarded time t_{ret} defined by

$$t_{ret} \equiv t - \frac{r}{c}. \tag{3.52}$$

Note that the definition of $r=0$ affects the definition of retarded natural modes and far natural modes. Ideally $r=0$ is chosen at or near the "center" of the antenna or scatterer. Appropriate choice of $r=0$ according to the symmetries of the object can carry the object symmetries over into retarded and far natural mode symmetries.

Retarded natural modes are a convenience for graphical or numerical display. Note that the growth of the natural modes as r increases does not cause a problem in the time domain if causality is enforced in the form of a delay for the signal to reach the observation position of interest (see also Subsect. 1.3.2).

3.4.3 Far Natural Modes

Far fields and potentials have been typically defined as r times the quantity of interest, with $r \to \infty$ and time converted to retarded time. For the fields these take the form of [3.26]

$$r\tilde{E}_f = -s\mu_0 \langle \tilde{\underline{g}}_{0_f}(r_0, r'; s); \tilde{J}(r', s) \rangle + \frac{s}{c} \langle \tilde{g}_{0_f}(r_0, r'; s) r_0 \times \tilde{J}_m(r', s) \rangle$$

$$r\tilde{H}_f = -\frac{s}{c} \langle \tilde{g}_{0_f}(r_0, r'; s) r_0 \times \tilde{J} \rangle - s\varepsilon_0 \langle \tilde{\underline{g}}_{0_f}(r_0, r'; s); \tilde{J}_m(r', s) \rangle \tag{3.53}$$

$$r\tilde{\underline{F}}_{q_f} = -s\mu_0 \underline{1} - r_0 r_0 + q i r_0 \times \underline{1} \cdot \langle \tilde{g}_{0_f}(r_0, r'; s), \tilde{K}_q(r', s) \rangle$$

where the far Green's functions are

$$\tilde{g}_{0_f}(r_0, r'; s) = \frac{e^{\gamma r_0 \cdot r'}}{4\pi}$$

$$\tilde{\underline{g}}_{0_f}(r_0, r'; s) = \tilde{g}_{0_f}(r_0, r'; s)[\underline{1} - r_0 r_0]. \tag{3.54}$$

The far natural modes for the fields are defined by

$$v_{f_\alpha}^{(X)}(r_0) \equiv \lim_{r \to \infty} \frac{r}{l_\alpha} v_{ret_\alpha}^{(X)}(r)$$

$$= \lim_{r \to \infty} \frac{r}{l_\alpha} e^{\gamma_\alpha r} v_\alpha^{(X)}(r) \tag{3.55}$$

where l_α is some appropriate length for a scaling constant. The far natural modes for the fields can be calculated from

$$v_{f_\alpha}^{(E)} = -\frac{c_\alpha \gamma_\alpha^2}{l_\alpha} \langle \tilde{g}_{0_{f_\alpha}}(r_0, r'; s); v_\alpha^{(J)}(r' s) \rangle$$

$$+ \frac{c_\alpha \eta_\alpha' \gamma_\alpha^2}{l_\alpha} \langle \tilde{g}_{0_{f_\alpha}}(r_0, r'; s) r_0 \times v_\alpha^{(Jm)}(r', s) \rangle$$

$$v_{f_\alpha}^{(H)} = -\frac{c_\alpha \gamma_\alpha^2}{l_\alpha} \langle \tilde{g}_{0_{f_\alpha}}(r_0, r'; s) r_0 \times v_\alpha^{(Jm)}(r', s) \rangle \tag{3.56}$$

$$- \frac{c_\alpha \eta_\alpha' \gamma_\alpha^2}{l_\alpha} \langle \tilde{g}_{0_{f_\alpha}}(r_0, r'; s); v_\alpha^{(Jm)}(r', s) \rangle$$

$$v_{f_\alpha}^{(Fq)} = -\frac{c_\alpha \gamma_\alpha^2}{l_\alpha} [\mathbf{1} - r_0 r_0 + q i r_0 \times \mathbf{1}] \cdot \langle \tilde{g}_{0_{f_\alpha}}(r_0 r'; s), v_\alpha^{(Kq)}(r', s) \rangle$$

where η_α' gives the relative strength of the magnetic current density natural mode if these occur.

To obtain the far field modes from the current density modes in Subsection 3.3.5 the additional coefficient for rF_q or rE is $Z_0 l_\alpha/(c_\alpha \gamma_\alpha)$ and for rH is $l_\alpha/(c_\alpha \gamma_\alpha)$. The expansions then become far-field expansions.

3.4.4 Numerical Examples

Examples of radiated fields are relatively new in the SEM literature. At present they are contained in one report by TESCHE [3.25]. The basic geometry considered so far is a finite length cylindrical antenna driven in the center, with a possible symmetrical impedance loading along the antenna as illustrated in Fig. 3.22. If the antenna is perfectly conducting (zero loading impedance), the far natural modes for the E field have normalized forms, as illustrated in Fig. 3.23. These modes are rather interesting in that they represent time-domain antenna patterns. They show that each complex frequency pole radiates in preferred directions, i.e., the pole residues are stronger in some directions than others. Equivalently a particular damped sinusoid yields larger transient fields in some directions than in other directions.

If a resistive loading of the form $C_3/(h-|z|)$ is added to the antenna, then some interesting things happen. As C_3 (real) is varied the lowest order pole pair coalesces at a critical damping value as illustrated in Fig. 3.24. This coalescing defines a critically damped antenna in that the lowest-order pole pair becomes a second-order pole for this special loading. The far E-field modes change somewhat due to the resistive loading, as shown in Fig. 3.25. Note the smoothing of the pattern for the higher-order modes in the sense of not-so-deep nulls (excluding $\theta = 0°$, $90°$).

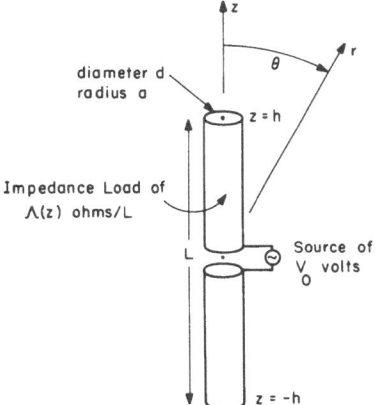

Fig. 3.22. Linear impedance loaded EMP simulator

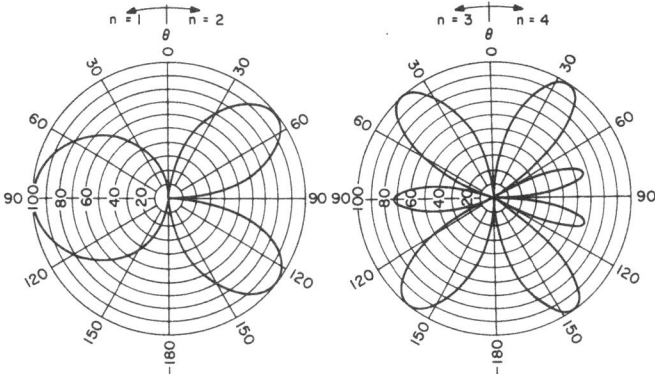

Fig. 3.23. Polar plots of the magnitude of the normalized far field modes for first layer of poles of an unloaded antenna. n denotes the individual poles in that layer

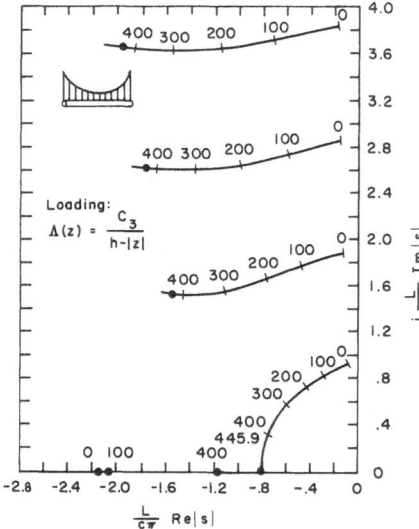

Fig. 3.24. Pole trajectories for resistively loaded wire of a $a/h=0.01$ as a function of the constant C_3 [kilo-ohms]

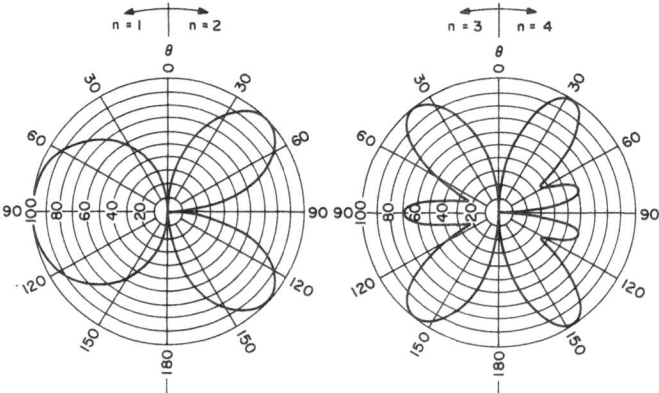

Fig. 3.25. Polar plots of the magnitude of the normalized far field modes for $l=1$ poles of the $(h-|z|)^{-1}$ resistively loaded antenna

Again, time history plots of the far field were made for step excitation, with excellent agreement for less than 20 included pole pairs. Class 1 coupling coefficients were used for both current and fields. Figure 3.26 shows the results for time-domain current and far field for $C_3/(h-|z|)$ loading, computed with seven pole pairs. In the frequency domain significant discrepancy was noted for the input admittance. It has since

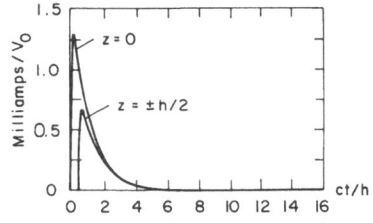

a) Time history plots of the current at the
input of the antenna and at z = ±h/2

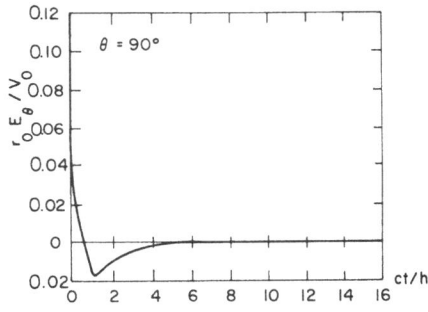

b) Time history plot of the radiated electric field

Fig. 3.26a and b. Step response of
the $(h-|z|)^{-1}$ resistively loaded
antenna, with $C_3 = 445.9\,\Omega$,
$a/h = 0.01$

been found [3.51] that the difference is an entire function of a simple type (a real constant) which can be computed from the admittance at $s=0$.

When the Green's functions are used to integrate over the current density to obtain the fields, there is another ambiguity in the definition of the field-natural-modes (including retarded and far field). The Green's function can be evaluated at s_α or left in a convolution form with a general s. This ambiguity, as with the classes 1 and 2 coupling coefficients, relates to the entire function contribution which must be evaluated for each type of problem. Both forms give the same correct pole residue. One form with its pole is the other plus an entire function.

3.5 Numerical Procedures

3.5.1 The Newton and Muller Methods for Finding Natural Frequencies

For finding the important pole terms in the singularity expansion, the first step is finding the natural frequencies. In an exact or approximate representation of the solution in terms of known functions, the calculations of the natural frequencies pose often a much simpler task as, for example, in the case of the perfectly conducting sphere. Such simpler cases rely on considerations peculiar to the problem at hand and are not considered here.

For a more general type of object a numerical MoM formulation is typically employed. Define

$$\tilde{\Delta}_N(s) \equiv \det((\tilde{\Gamma}_{n,m}(s))) \tag{3.57}$$

where $(\tilde{\Gamma}_{n,m}(s))$ is the $N \times N$ matrix approximation of the integral equation operator, as discussed in Subsection 3.2.2. The natural frequencies satisfy the equation

$$\tilde{\Delta}_N(s_\alpha) = 0. \tag{3.58}$$

One can then find the natural frequencies by searching for the roots of $\tilde{\Delta}_N(s)$, using its properties as an analytic function of the complex frequency.

Various root finding procedures can now be employed. Some common ones are the Newton and Muller methods. Sufficiently near a particular s_α, one can make a linear approximation to the function by the evaluation of $\tilde{\Delta}_N$ and its derivative with respect to s, or by the evaluation of $\tilde{\Delta}_N$ at two values of s. Setting the linear approximation to zero updates one's initial guess at s_α to a new approximation for s_α. Repeating this process until it converges to within some particular accuracy of a root is the essence of Newton's method. This process can be repeated to find other roots. As each s_α is found, $\tilde{\Delta}_N(s)$ can be divided by $s - s_\alpha$ to remove the already found roots (for first-order zeros of $\tilde{\Delta}_N(s)$) from further consideration.

Going a step further one can make a parabolic fit to $\tilde{\Delta}_N$. This requires the value of $\tilde{\Delta}_N$ at three points, or two points and a derivative, etc. The roots of the parabolic fit are used to improve the computed value of s_α and the process is repeated. This procedure is called Muller's method. Constructing the parabola from values of $\tilde{\Delta}_N$ for three values of s has been often used in SEM studies and is described by CROW et al. [3.29]. In order to insure that all the poles in a particular region of the s plane have been found, one can plot contours of constant $|\tilde{\Delta}_N(s)|$. With a sufficiently large number of such contours, the s_α are surrounded by concentric circles (for small radii) and thus readily identified.

Since the determinant is zero if at least one of the eigenvalues is zero (with the matrix elements bounded), one can look for such eigenvalue zeros. In matrix triangularization techniques the diagonal elements are related to the eigenvalues. If one diagonal element in a product of triangular matrices is zero then the determinant is zero and an s_α is found. One can then use Newton's or Muller's method on some small diagonal term in a triangularized matrix derived from $(\tilde{\Gamma}_{n,m}(s))$. This type of search for the s_α has also been used in SEM studies employing MoM numerical formulations.

3.5.2 Contour Integral Methods

There has been some recent investigation into the numerical implementation of certain types of contour integrals to determine whether there are s_α in some particular region of the s plane, and to approximately calculate the s_α [3.31]. If $\tilde{\Delta}_N(s)$ is the function of interest then the argument number can be defined as

$$
\begin{aligned}
N_a &\equiv \frac{1}{2\pi i} \oint_C \frac{1}{\tilde{\Delta}_N(s')} \frac{d}{ds'} \tilde{\Delta}_N(s')\,ds' \\
&= \frac{1}{2\pi} \arg(\tilde{\Delta}_N(s))\Big|_C
\end{aligned}
$$

$$
\begin{aligned}
=\,&(\text{number of zeros inside C, including multiplicity}) \\
&-(\text{number of poles inside C, including multiplicity}) \qquad (3.59)
\end{aligned}
$$

where C is a simple closed counterclockwise contour in the s plane. This argument number can be readily determined from the change in the argument (phase) of the function around the contour. Typically, only zeros are present in regions of the s plane of interest so evaluation of N_a is a way of determining how many s_α are to be found in particular regions of the s plane. Since the s_α are singularities of the response, N_a can be appropriately referred to as a singularity number. Under parameter variation, the s_α move and N_a can be used to define a conservation of singularities concept.

The Cauchy integral formula can be used to define the function values inside C from those on C from

$$
\frac{d^n}{ds^n} \tilde{\Delta}_N(s) = \frac{n!}{2\pi i} \oint_C \frac{\tilde{\Delta}_N(s')}{(s'-s)^{n+1}}\,ds' \qquad n=0,1,2,3,\ldots \qquad (3.60)
$$

This is numerically interesting in that one can calculate a set of values of $\tilde{\Delta}_N$ on the contour and store them for subsequent use in finding $\tilde{\Delta}_N(s)$ and its derivatives inside C, which can be used in Newton's or

Muller's method for finding the s_α inside C. Since determinants and other matrix operations can often be time consuming on computers, one would prefer not to have to calculate them repeatedly in an iterative process.

Another formula of interest is

$$\frac{1}{2\pi i} \oint_C \tilde{g}(s') \frac{1}{\tilde{\Delta}_N(s')} \frac{d}{ds'} \tilde{\Delta}_N(s')ds'$$

$$= \sum \tilde{g}(s) \quad \text{(at zeros of } \tilde{\Delta}_N \text{ inside C)}$$
$$- \sum \tilde{g}(s) \quad \text{(at poles of } \tilde{\Delta}_N \text{ inside C)} \qquad (3.61)$$

with multiplicities counted. Here $\tilde{g}(s)$ is assumed to be an analytic function inside C. If $\tilde{\Delta}_N(s)$ is analytic within and on C (and thus has no poles inside C) then (3.59) can be used to ascertain whether there is only one first-order zero inside C corresponding to $N_a = 1$. If this is the case then setting $\tilde{g}(s) \equiv s$ we have

$$s_\alpha = \frac{1}{2\pi i} \oint_C s' \frac{1}{\tilde{\Delta}_N(s')} \frac{d}{ds'} \tilde{\Delta}_N(s')ds' \qquad (3.62)$$

as an explicit formula for a natural frequency from the values of the determinant on the contour.

Other types of numerical contour integrals can perhaps also be used for finding the s_α. This subject area requires more investigation and practical implementation in order to speed numerical computation and to expand the kinds of parameters that can be profitably defined and used in connection with the s_α.

3.5.3 Pole Tracking in Parameter Variation

One of the important characteristics of the pole terms is the manner in which different factors contain information about various aspects of the electromagnetic problem. The variation of these factors provides a useful way to study the dependence of the response on the geometric shape, impedance loading, etc.

For the natural frequencies consider the equation

$$\tilde{\Delta}_N(s_\alpha, P) = 0 \qquad (3.63)$$

where P is some parameter of interest and $\tilde{\Delta}_N$ is the determinant or an appropriate related function. Then one has an equation for the change of the natural frequencies as

$$\frac{d s_\alpha}{d P} = - \frac{\dfrac{\partial \tilde{\Lambda}_N}{\partial P}}{\dfrac{\partial \tilde{\Lambda}_N}{\partial s_\alpha}}.$$ (3.64)

Given some starting values of s_α, P, the above equation can be numerically integrated to find s_α as a function of P.

This technique has been recently proposed by WILTON et al. [3.32] and tested with good results on the perfectly conducting loop, with P as the wire radius.

3.5.4 Natural Modes and Related Quantities

Having found a particular s_α and having determined the pole order, one can proceed to calculate the natural mode(s) and related quantities in a numerical MoM formulation. Assuming a first-order pole without modal degeneracy the numerical procedure is straightforward. One must solve

$$(\tilde{\Gamma}_{n,m}(s_\alpha)) \cdot (v_n)_\alpha = (0_n) \qquad n, m = 1, 2, 3, \dots, N.$$ (3.65)

One can do this by arbitrarily setting one component of $(v_n)_\alpha$ equal to some non zero number, removing this component to deflate $(v_n)_\alpha$ and $(\Gamma_{n,m}(s_\alpha))$ to $N-1$ size, producing a non-zero right hand vector, and inverting the $(N-1) \times (N-1)$ matrix. The natural mode can then be appropriately normalized. Of course, there are precautions to be taken such as avoiding a null in the natural mode in the choice of a starting element. If $(\tilde{\Gamma}_{n,m}(s_\alpha))$ is not symmetric, the numerical coupling vector (the left-side vector) $(\mu_n)_\alpha$ can be determined by the same procedure as above. The coupling coefficient then requires some matrix and vector multiplication which in the MoM form is the equivalent of the integration formulas given in Section 3.3.

3.6 Contemporary Developments

3.6.1 Equivalent Circuits

Someone with electrical engineering experience can note the similarity of some of the SEM concepts and circuit theory concepts. Circuit theory should be able to provide many insights into future SEM developments.

To each pole or pole pair in the SEM response, one can associate an LRC resonant circuit. In particular, for an antenna of finite size in free space, with a single terminal pair, one can characterize the short circuit current and the admittance by the same poles (and similarly for the open circuit voltage and the impedance). This allows one to have tank circuits with sources to establish the proper amplitude (coupling coefficient) for each pole. This author and TESCHE [3.51] have some results on this problem which will hopefully be reported in the near future.

3.6.2 Eigenmodes and Their Relation to SEM

There has been some work by GARBACZ [3.17], TURPIN [3.18], and HARRINGTON and MAUTZ [3.20] concerning what they refer to as characteristic modes with certain convenient power orthogonality properties. Another related type of mode which will be referred to as eigenmodes can also be defined, with certain useful properties in an SEM (s plane) sense [3.21]. Using the general integral equation form in Section 3.3 these are defined by

$$\langle \tilde{\underline{\Gamma}}(r,r';s), \tilde{R}_\beta(r',s) \rangle = \tilde{\lambda}_\beta(s) \tilde{R}_\beta(r,s)$$
$$\langle \tilde{L}_\beta(r,s), \tilde{\underline{\Gamma}}(r,r';s) \rangle = \tilde{\lambda}_\beta(s) \tilde{L}_\beta(r,s).$$

$$(3.66)$$

These can be used to represent the current on an object and can be extended to the radiated or scattered fields. Note that eigenvalues and eigenmodes are concepts basically different from natural frequencies and natural modes.

Since the zeros of the eigenvalues $\tilde{\lambda}_\beta(s)$ are the s_α the analytic properties of the eigenvalues are of considerable interest. Under certain restrictions the separate eigenvalues (different β) each have some of the s_α as their zeros, thus splitting the s_α into different groups.

For certain definitions of the operator $\tilde{\Gamma}(s)$ the eigenvalues can be thought of as eigenimpedances. Such an operator relates a response current density to a source electric field. This type of operator (or matrix) is referred to as an impedance operator (or matrix) [3.47]. For certain general forms of loading impedances, including lumped and distributed, the eigenvalues after loading are equal to the eigenvalues before loading plus the impedance, i.e., we have the general transformation

$$\tilde{\lambda}_\beta(s) \to \tilde{\lambda}_\beta(s) + \tilde{Z}(s).$$

$$(3.67)$$

As impedances the eigenvalues have interesting s plane properties. This introduces the possibility of antenna and scatterer synthesis in a form similar to circuit synthesis. If one plots the eigenvalues in the s plane (contours of magnitude and phase) then the zeros of the $\tilde{\lambda}_\beta(s)$ can be shifted by the addition of a judiciously chosen $\tilde{Z}(s)$. This is basically a root locus type of synthesis technique for choosing some "best" $\tilde{Z}(s)$, which should prove to be a potent design technique for transient and broadband antennas such as EMP simulators.

3.6.3 SEM Analysis of Experimental Data

As mentioned before, the singularity expansion method was suggested from observation of damped sinusoids in the electromagnetic response of various complicated scatterers [3.2]. SEM is then not confined to being a part of the mathematical process in solving electromagnetic boundary value problems; it can be used to organize experimental data as well.

Experimental transient data is in general Laplace transformable and thereby has singularities in the complex frequency plane. These singularities can be used to characterize the data and manipulate it into other forms (such as frequency response). For cases where one expects only poles in the finite s plane, this has been tried with some success by MITTRA (see Sect. 2.6). The same data analysis techniques for transient experimental data can also be used to analyze calculated transient waveforms for the SEM quantities. Such calculated waveforms can be obtained from any of a number of techniques. It may be possible to apply certain filtering concepts to remove some error or noise by only using those SEM terms which are physically allowed (passive, causal, etc.). One should also be able to take cw response information and analytically continue it into the complex plane to find the singularities. Much work is needed in this data analysis area, and error analysis will most likely be quite important. The data analysis problem can also be thought of as a target response experimental analysis and hence a target recognition (inverse scattering) procedure.

3.7 Summary

The more one thinks about the singularity expansion method the simpler the basic concept becomes. The basic idea of SEM is to express electromagnetic behavior in terms of the singularities in the complex frequency plane. Actually, SEM is also applicable to other types of physical problems such as acoustics, mechanics, etc.

Most SEM work to date has been concentrated on the calculation of the response of finite size objects in free space such as antennas and scatterers. For this case, only poles appear in the finite s plane, giving rise to a considerable simplification. Other kinds of objects, as well as propagation problems, need to be considered to extend the understanding and practical utility of SEM. More than natural frequencies, natural modes, and coupling coefficients will be involved. Nonlinear problems need consideration as well. It has been found that SEM can be readily combined with MoM for treating the response of rather general types of objects. One should not lose sight, however, of the fact that SEM is inherently neither a numerical procedure nor an analytic formula. SEM is a representation in terms of the s plane singularities in whatever type of problem formulation is used.

While important results have been obtained so far, we have perhaps only scratched the surface of a larger body of results yet ot be obtained. To name a few obvious directions of study, there are the equivalent circuits; use of, and relation to, eigenmodes; antenna and scatterer synthesis; analysis of experimental data; and target recognition (inverse scattering). In addition there are certain types of problems for which branch cuts as well as poles are needed in the complex frequency plane. General formalisms for handling such branch cuts are required to improve physical understanding and simplify the practical use of the results.

Acknowledgements

This author would like to thank the many investigators who have contributed to the development of the theory and practice of SEM. They are identified by references and discussion. Being privileged to be personally acquainted with almost all of them, this author wishes to thank especially those people who helped him get SEM off the ground when but a few years ago SEM was a fresh idea in his mind: Drs. L. MARIN, K. S. H. LEE, R. LATHAM, and F. TESCHE.

References

3.1   C. E. BAUM: "Interaction of Electromagnetic Fields with an Object which Has an Electromagnetic Symmetry Plane", Interaction Note 63, March 1971

3.2   C. E. BAUM: "On the Singularity Expansion Method for the Solution of Electromagnetic Interaction Problems", Interaction Note 88, Dec. 1971

3.3   L. MARIN, R. W. LATHAM: "Analytical Properties of the Field Scattered by a Perfectly Conducting, Finite Body", Interaction Note 92, Jan. 1972; also
      L. MARIN, R. W. LATHAM: Proc. IEEE **60**, 640 (1972);
      L. MARIN: IEEE Trans. Ap-**21**, 809 (1973)

3.4   L. MARIN: "Natural-Mode Representation of Transient Scattering from Rotationally Symmetric, Perfectly Conducting Bodies and Numerical Results for a Prolate Spheroid", Interaction Note 119, Sept.1972; also
      L. MARIN: IEEE Trans. AP-**22**, 266 (1974)

3.5   F. M. TESCHE: "On the Singularity Expansion Method as Applied to Electromagnetic Scattering from Thin Wires", Interaction Note 102, April 1972; also
      F. M. TESCHE: IEEE Trans. AP-**21**, 53 (1973)

3.6   J. P. MARTINEZ, Z. L. PINE, F. M. TESCHE: "Numerical Results of the Singularity Expansion Method as Applied to a Plane Wave Incident on a Perfectly Conducting Sphere", Interaction Note 112, May 1972

3.7   L. MARIN: "Application of the Singularity Expansion Method to Scattering from Imperfectly Conducting Bodies and Perfectly Conducting Bodies Within a Parallel Plate Region", Interaction Note 116, June 1972

3.8   S. W. LEE, B. LEUNG: "The Natural Resonance Frequency of a Thin Cylinder and Its Application to EMP Studies", Interaction Note 96, Feb. 1972

3.9   C. E. BAUM: "On the Singularity Expansion Method for the Case of First Order Poles", Interaction Note 129, Oct. 1972

3.10 C. D. TAYLOR, T. T. CROW, K. T. CHEN: "On the Singularity Expansion Method Applied to Aperture Penetration: Part I, Theory", Interaction Note 134, May 1973

3.11 P. R. BARNES: "On the Singularity Expansion Method as Applied to the EMP Analysis and Simulation of the Cylindrical Dipole Antenna", Interaction Note 146, Nov. 1973

3.12 F. M. TESCHE: "Numerical Considerations for the Calculation of Currents Induced on Intersecting Wires Using the Pocklington Integro-Differential Equation", Interaction Note 150, Jan. 1974

3.13 D. R. WILTON, K. R. UMASHANKAR: "Parametric Study of an L-Shaped Wire Using the Singularity Expansion Method", Interaction Note 152, Nov. 1973

3.14 P. R. BARNES, D. B. NELSON: "Transient Response of Low Frequency Vertical Antennas to High Altitude Nuclear Electromagnetic Pulse (EMP)", Interaction Note 160, March 1974

3.15 T. T. CROW, B. D. GRAVES, C. D. TAYLOR: "The Singularity Expansion Method as Applied to Perpendicular Crossed Cylinders in Free Space", Interaction Note 161, Oct. 1973

3.16 K. R. UMASHANKAR, D. R. WILTON: "Analysis of an L-Shaped Wire Over a Conducting Ground Plane Using the Singularity Expansion Method", Interaction Note 174, March 1974

3.17 R. J. GARBACZ: "A Generalized Expansion for Radiated and Scattered Fields", Interaction Note 180, 1968 (PhD dissertation, Ohio State University)

3.18 R. H. TURPIN: "Basis Transformation, Least Square, and Characteristic Mode Techniques for Thin-Wire Scattering Analysis", Interaction Note 181, 1970 (PhD dissertation, Ohio State University); also
      R. J. GARBACZ, R. H. TURPIN: IEEE Trans. AP-**19**, 348 (1971)

3.19 L. MARIN: "Natural Modes of Certain Thin-Wire Structures", Interaction Note 186, Aug. 1974

3.20 R. F. HARRINGTON, J. R. MAUTZ: "Theory and Computation of Characteristic Modes for Conducting Bodies", Interaction Note 195, Dec. 1970; also
      R. F. HARRINGTON, J. R. MAUTZ: IEEE Trans. AP-**19**, 622, 629 (1971)

3.21 C. E. Baum: "On the Eigenmode Expansion Method for Electromagnetic Scattering and Antenna Problems, Part I: Some Basic Relations for Eigenmode Expansions and Their Relation to the Singularity Expansion", Interaction Note 229, January 1975

3.22 R. H. Shafer, R. G. Kouyoumjian: "Transient Currents on a Cylinder Illuminated by an Impulsive Plane Wave", Interaction Note to be published and IEE Trans. AP to appear

3.23 L. Marin: "Transient Electromagnetic Properties of Two Parallel Wires", Sensor and Simulation Note 173, March 1973

3.24 L. Marin: "Natural Modes of Two Collinear Cylinders", Sensor and Simulation Note 176, May 1973

3.25 F. M. Tesche: "Application of the Singularity Expansion Method to the Analysis of Impedance Loaded Linear Antennas", Sensor and Simulation Note 177, May 1973

3.26 C. E. Baum: "Singularity Expansion of Electromagnetic Fields and Potentials Radiated from Antennas or Scattered from Objects in Free Space", Sensor and Simulation Note 179, May 1973

3.27 T. H. Shumpert: "EMP Interaction with a Thin Cylinder above a Ground Plane Using the Singularity Expansion Method", Sensor and Simulation Note 182, June 1973

3.28 K. S. H. Lee, L. Marin: "SGEMP for Resonant Structures", Theoretical Note 199, Sept. 1974

3.29 T. T. Crow, B. D. Graves, C. D. Taylor: "Numerical Techniques Useful in the Singularity Expansion Method as Applied to Electromagnetic Interaction Problems", Mathematics Note 27, Dec. 1972

3.30 C. E. Baum: "Electromagnetic Reciprocity and Energy Theorems for Free Space Including Sources Generalized to Numerous Theorems, to Combined Fields, and to Complex Frequency Domain", Mathematics Note 33, Dec. 1973

3.31 C. E. Baum: "On the Use of Contour Integration for Finding Poles, Zeros, Saddles, and Other Function Values in the Singularity Expansion Method", Mathematics Note 35, Feb. 1974

3.32 D. R. Wilton, R. J. Pogorzelski, R. D. Nevels: "Singularity Trajectories Under Parameter Variation", Mathematics Note (to be published)

3.33 J. Schwinger: "The Theory of Obstacles in Resonant Cavities and Waveguides", MIT Radiation Laboratory report (May 1943)

3.34 H. C. Pocklington: Proc. Cambridge Phil. Soc. 9, 324 (1897)

3.35 M. Abraham: Ann. Phys. (Leipzig) 66, 435 (1898)

3.36 Lord Rayleigh: Proc. Roy. Soc. 87, 193 (1912)

3.37 C. W. Oseen: Phys. Z. 14, 1222 (1913)
C. W. Oseen: Arkiv Mat. Astr. Fysik 9, no. 28, 1 (1914)
C. W. Oseen: Arkiv Mat. Astr. Fysik 9, no. 29, 1 (1914)
C. W. Oseen: Arkiv Mat. Astr. Fysik 9, no. 30, 1 (1914)

3.38 E. Hallén: "Über die elektrischen Schwingungen in drahtförmigen Leitern", Uppsala Univ. Årsskrift, no. 1, pp. 1–102, 1930;
E. Hallén: Nova Acta Reg. Soc. Sci. Upsaliensis 11, 1 (1938)

3.39 L. Page, N. Adams: Phys. Rev. 53, 819 (1938)
L. Page: Phys. Rev. 65, 98, 111 (1944)

3.40 E. T. Copson: Theory of Functions of a Complex Variable (Oxford Press, Oxford 1935)

3.41 J. A. Stratton: Electromagnetic Theory (McGraw Hill, New York 1941)

3.42 S. A. Schelkunoff, H. T. Friis: Antennas Theory and Practice, (Wiley, New York 1952)

3.43 R. V. CHURCHILL: *Complex Variables and Applications* (McGraw Hill, New York 1960)

3.44 J. VAN BLADEL: *Electromagnetic Fields* (McGraw Hill, New York 1964)

3.45 P. ROMAN: *Advanced Quantum Theory* (Addison-Wesley, New York 1965)

3.46 G. CARRIER, M. KROOK, C. PEARSON: *Functions of a Complex Variable* (McGraw Hill, New York 1966)

3.47 R. F. HARRINGTON: *Field Computation by Moment Methods* (Macmillan, 1968); also for recent treatments
R. MITTRA (ed.): *Topics in Applied Physics*, Vol. 3: Numerical and Asymptotic Techniques in Electromagnetics (Springer, Berlin, Heidelberg, New York 1975)

3.48 L. A. WEINSTEIN: *Open Resonators and Open Waveguides* (Golem Press, Boulder, Colo. 1969)

3.49 C. T. TAI: *Dyadic Green's Functions in Electromagnetic Theory* (Intext Educational Publishers, 1971)

3.50 L. B. FELSEN, N. MARCUVITZ: *Radiation and Scattering of Waves* (Prentice-Hall, Englewood Cliffs, N. J. 1973)

3.51 F. M. TESCHE: Private communication

4. Radiation and Reception of Transients by Linear Antennas

D. L. Sengupta and C.-T. Tai

With 23 Figures

In this chapter we consider the characteristics of transient radiation and reception of linear antennas. The antennas treated are perfectly conducting cylinders excited symmetrically at the centers by slice generators having time-dependent input signals. We also consider linear antennas loaded with resistive material such that the internal resistance per unit length of the antenna varies in some predetermined fashion. Such loaded antennas may have application in reducing the effects of reflections from the antenna end points and also in shaping the wave forms of the radiated pulses. Some of the analytical and numerical techniques used to study the transients in linear antennas are reviewed and discussed. A few examples and some results are given for illustrative purposes only. Detailed discussions of various analytical methods and results may be found in the references cited.

A general discussion of radiation and reception of transients by linear antennas is given in Section 4.2. Classical solutions of transient problems are based on Fourier (or Laplace) transform techniques. The Fourier transform method of solving transient problems in linear antennas and the numerical solutions of such problems based on similar techniques are reviewed in Section 4.3. Section 4.4 considers the analytical solution based on application of the transform technique to wave forms radiated by a Gaussian-pulse-excited reflectionless linear antenna. As an example of the numerical method based on Fourier transform techniques, the transient radiation from a step-voltage-excited reflectionless antenna is considered in Section 4.5. Modern numerical techniques in transient electromagnetics use numerical methods directly in the time domain. These methods are mainly based on the solutions of space-time integral equations of the current distribution. The space-time integro-differential equation for the current distribution in a linear antenna and its numerical solution are discussed in Section 4.6. The method of solving a Hallen type of integral equation in the time domain is discussed in Section 4.7. Section 4.8 deals briefly with numerical solutions directly in the time domain without the use of space-time integral equations. Rigorous analytical solutions of infinitely long linear antennas are discussed in Section 4.9. The most significant omission from the present chapter is a discussion of the singularity expansion method (SEM) applied to linear antennas, which is covered in Chapter 3.

4.1 Background Material

The transient field of a linear antenna was apparently first investigated in [4.1] by using a linear current element as the source of radiation. Subsequently this work has been reviewed and reformulated qualitatively in [4.2]. There is no mention in these works as to how such a current pulse is excited. A scalar version of the problem has been studied in [4.3–6]. Transient transmission and reception properties of thin cylindrical dipoles excited by step voltage or rectangular pulses of various durations have been studied both theoretically and experimentally in [4.7–10]. Theoretical results given in [4.7–10] are of limited value because of the fact that the frequency domain results were prematurely terminated. Transmission and reception of reactangular pulse signals by wide angle biconical antennas have been investigated in [4.11] by use of an approximate input impedance function for the antenna and neglect of the interaction of the higher-order modes which exist within the conical region of the antenna. The transient response of a receiving linear antenna has also been investigated both theoretically and experimentally in [4.12]. Although the theoretical treatment in [4.12] is not based on the conventional method of treating the problem as an initial value problem, it is found useful to interpret the experimental results obtained. All the results mentioned so far tend to show in general that for monopoles of length very short compared to the wavelength of the highest significant frequency in the input signal, the radiated field is proportional to the second derivative in time of the input signal; when the short monopole is receiving, the received signal is proportional to the first time derivative of the incident field. These differences between transmission and reception of transient signals by linear antennas agree with the results apparently first predicted in [4.5]. If a simple integrating circuit is used for a load, the short receiving monopole will act as a faithful receiver of transient signals.

In general, the mathematical analysis of the transient radiation from linear antennas comes under the domain of initial boundary value problems. Rigorous analytical solutions of such time-dependent boundary value problems are extremely difficult, and are available only for some highly restricted cases. Exact theoretical expressions have been derived in [4.13,14] for the transient current distribution on an unloaded infinitely long linear antenna excited by a step voltage across an infinitesimal center gap. The more general problem of obtaining the fields anywhere in space for a step-voltage-excited unloaded dipole is treated theoretically in [4.15]. Some numerical results on the current distribution based on [4.15] are given in [4.16]. Transient radiation from unloaded and uniformly loaded infinitely long linear antennas excited with step voltage is discussed theoretically in [4.17]. Analytical investigations of the waveforms radiated by a non-uniformly loaded finite linear antenna excited by Gaussian input signal are discussed in [4.18].

Numerical solutions based on Fourier transform techniques are given in [4.19] for the transient waveforms radiated by a Gaussian-pulse-excited infinitely long linear antenna, in [4.20, 21] for the transient fields produced by thin cylindrical antennas driven by dc pulses and in [4.22] for step-voltage-excited non-uniformly loaded finite linear antennas. Some results have been obtained [4.23–25] for the transient waveforms radiated by both unloaded and uniformly loaded finite linear antennas by applying numerical methods in the time domain via the space-time integral equation for the current distribution in the antenna. Wave forms radiated by step-voltage-excited dipole antennas have been obtained in [4.26] by applying moment methods directly in the time domain. An excellent summary of the computational aspects of transient electromagnetics in general may be found in [4.27] (see also Chapt. 2). The application of the singularity expansion method to obtain transient radiation from linear antennas has been discussed in [4.28] (see also Chapt. 3).

4.2 Transient Radiation and Reception

In order to clarify some of the fundamental antenna concepts involved, we present in this section a general formulation of the problem of transient radiation and reception by linear antennas. Some simple examples are given to illustrate the underlying principles which govern the transfer of signals from a transmitting to a receiving antenna. It will be shown that the transfer function between the two antennas can drastically modify the wave form of the transmitted field. Since the ordinary definition of antenna directivity for time-harmonic excitation is no longer applicable for transient fields, we propose here a new definition for the directivity of an antenna radiating a transient signal. The usefulness of the proposed antenna parameter has yet to be tested and ascertained.

4.2.1 The Transmitting Antenna

The Fourier transform of a time function such as $E(r,t)$ or $E(t)$ will be denoted by $E(r,\omega)$ or $E(\omega)$ so that

$$E(r,\omega) = \int_{-\infty}^{\infty} E(r,t)\exp(i\omega t)dt\,, \tag{4.1}$$

and

$$E(r,t) = \frac{1}{2\pi} \int_{-\infty}^{\infty} E(r,\omega)\exp(-i\omega t)d\omega\,, \tag{4.2}$$

where r denotes the position vector of the field point (or the observation point) with respect to a given origin. With the above definition of the Fourier transform relations, the Maxwell equations in the frequency domain assume the same form as those of a harmonically oscillating field with the time convention $\exp(-i\omega t)$, namely,

$$\nabla \times E(r,\omega) = i\omega\mu_0 H(r,\omega) - J_m(r,\omega)\,, \tag{4.3}$$

$$\nabla \times H(r,\omega) = -i\omega\varepsilon_0 E(r,\omega) + J(r,\omega)\,, \tag{4.4}$$

where μ_0, ε_0 are the permeability and permittivity of free space, $E(r,\omega)$, $H(r,\omega)$ are the time harmonic electric and magnetic field vectors at r, and $J(r,\omega)$, $J_m(r,\omega)$ are the time harmonic electric and magnetic source current density vectors.

The important parameters describing the essential characteristics of a transmitting antenna for harmonically oscillating fields are:

a) the input impedance, denoted by $Z_i(\omega)$

b) the vector effective height function, denoted by $h(\omega)$. The effective height of a transmitting linear antenna is defined as the moment of its source current distribution divided by the input current. For a receiving linear antenna, the effective height is defined as the voltage induced between the open terminals of the antenna divided by the incident electric field intensity when the field is parallel to the antenna. For a given antenna these two effective heights are equal [4.29]. Note that both $Z_i(\omega)$ and $h(\omega)$ are in general functions of ω.

In terms of the vector effective height function the time independent far-zone electric field produced by a transmitting antenna can be written as follows

$$E(r,\omega) = [i\omega\eta_0 I_i(\omega) h(\omega)\exp(i\omega r/c)]/4\pi cr\,, \tag{4.5}$$

where $\eta_0 = (\mu_0/\varepsilon_0)^{1/2}$ is the intrinsic impedance of free space, $c = (\mu_0\varepsilon_0)^{-1/2}$ is the velocity of light in free space, r is the distance measured from a reference origin to the observation point where the electric field is observed, and $I_i(\omega)$ denotes the input current to the antenna at a well-defined pair of terminals. The term well-defined pair of terminals implies that we can define an input impedance of the antenna uniquely and unambiguously.

For the purpose of evaluating the time independent far field quantities explicitly it is found convenient to express the vector effective height of the antenna in terms of Schelkunoff's radiation vectors N and M [4.29]. This is done through the following relations

$$h(\omega) = (1/I_i)(N_t - r_0 \times M_t), \tag{4.6}$$

where

$$N_t = N - N_r r_0, \tag{4.7}$$

$$M_t = M - M_r r_0, \tag{4.8}$$

and

$$N = \int_{V'} J(r',\omega) \exp\{-i\omega c^{-1}[(r'\cdot r_0)]\} dv', \tag{4.9}$$

$$M = \frac{1}{\eta_0} \int_{V'} J_m(r',\omega) \exp\{-i\omega c^{-1}[(r'\cdot r_0)]\} dv', \tag{4.10}$$

where r' is the position vector of the source point, V' is the volume containing the source currents, and r_0 is the unit vector in the direction of the field point, i.e., in the direction of r. The vector effective height defined here differs by a sign from the one originally introduced in [4.30]. The input current to an antenna is related to the input voltage $V_i(\omega)$ by the relation

$$I_i(\omega) = V_i(\omega)/Z_i(\omega), \tag{4.11}$$

where $V_i(\omega)$ is the Fourier spectrum of the input voltage.

Based on (4.5) we can write the time domain far-zone electric field due to an input current or input voltage as

$$E(r,t) = \frac{i\eta_0}{(8\pi^2 cr)} \int_{-\infty}^{\infty} \omega I_i(\omega) h(\omega) \exp\{-i\omega[t - (r/c)]\} d\omega \tag{4.12}$$

$$= \frac{i\eta_0}{(8\pi^2 cr)} \int_{-\infty}^{\infty} [\omega V_i(\omega)/Z_i(\omega)] h(\omega) \exp\{-i\omega[t - (r/c)]\} d\omega.$$

To illustrate the significance of (4.12), let us consider the case of a center driven linear traveling wave antenna of length $2L$ aligned along the z-axis as shown in Fig. 4.1 and carrying a time harmonic current represented by

$$I(z,\omega)=I_0(\omega)\exp\left[(i\omega/c)|z|\right], \qquad -L\leq z\leq L. \qquad (4.13)$$

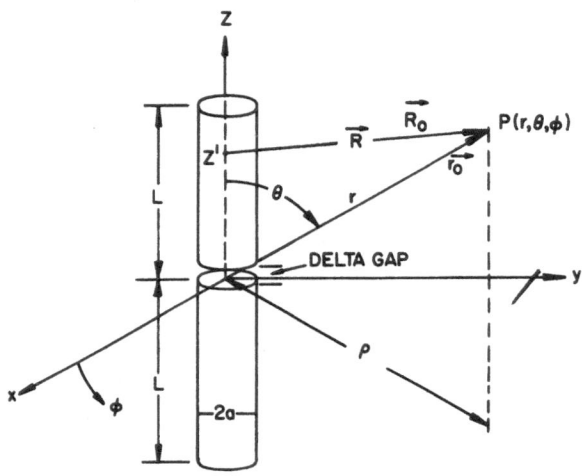

Fig. 4.1. A linear antenna with a slice generator at the delta gap. The various coordinate systems used are also shown

After using the appropriate relations given above, it can be shown that the effective height of the antenna is given by

$$\boldsymbol{h}(\omega)=ic\omega^{-1}\sin\theta\left\{2/\sin^2\theta-\frac{\exp\left[(i\omega L/c)(1+\cos\theta)\right]}{(1+\cos\theta)}\right.$$
$$\left.-\frac{\exp\left[(i\omega L/c)(1-\cos\theta)\right]}{(1-\cos\theta)}\right\}\boldsymbol{\theta}_0 \qquad (4.14)$$

where θ denotes the angle of the field point as measured from the axis of the linear antenna, and $\boldsymbol{\theta}_0$ is the unit vector in the direction of θ. Using (4.14) and (4.12) we obtain

$$E(r,t) = \boldsymbol{\theta}_0 \eta_0 (8\pi^2 r)^{-1} \sin\theta \int_{-\infty}^{\infty} I_0(\omega)\{(2/\sin^2\theta)$$

$$-(1+\cos\theta)^{-1}\exp[i(\omega L/c)(1+\cos\theta)]$$

$$-(1-\cos\theta)^{-1}\exp[i(\omega L/c)(1-\cos\theta)]\}\exp[-i\omega(t-r/c)]\,d\omega$$

$$= \boldsymbol{\theta}_0 \eta_0 (4\pi r)^{-1} \sin\theta \{(2/\sin^2\theta)I_0(t-r/c) \qquad (4.15)$$

$$-(1+\cos\theta)^{-1}I_0[t-(r/c)-(L/c)(1+\cos\theta)]$$

$$-(1-\cos\theta)^{-1}I_0[t-(r/c)-(L/c)(1-\cos\theta)]\}$$

where $I_0(t)$ represents the input time domain current given by

$$I_0(t) = \frac{1}{2\pi} \int_{-\infty}^{\infty} I_0(\omega)\exp(-i\omega t)\,d\omega. \qquad (4.16)$$

The three terms in (4.15) represent the contributions to the field due to three wavelets originating from the center, the lower, and upper endpoints of the antenna, respectively. The significance of various terms contained in (4.15) is illustrated graphically in Fig. 4.2 where the origin of each wavelet is shown and the resultant field, as a function of t, is plotted in the right-hand side of the figure. When L/c is small compared to the pulse duration of the signal, (4.15) may be approximated by

$$E(r,t) \simeq (\eta_0 \sin\theta/2\pi r)(L/c)[\partial I(t-r/c)/\partial t]\boldsymbol{\theta}_0. \qquad (4.17)$$

This result is compatible with the result obtained in [4.3, 6].

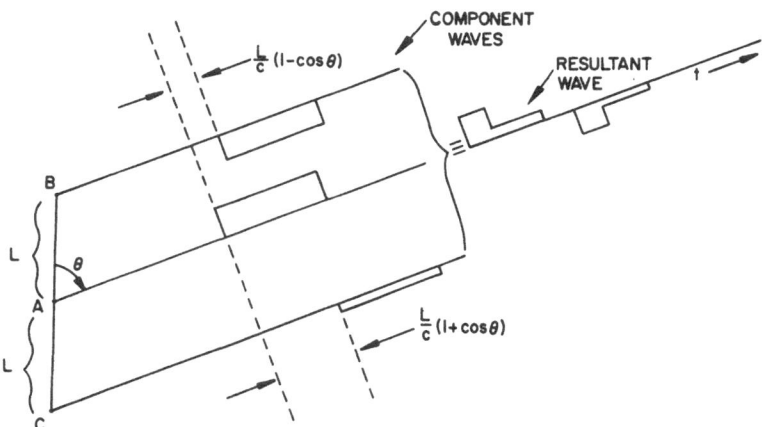

Fig. 4.2. The transient field of a center-driven linear antenna supporting a traveling current

The assumption of a purely traveling current wave on an unloaded linear antenna is physically unrealistic because there is no device which can be placed at the ends of such a linear antenna in free space to eliminate end reflections. A more realistic model, therefore, should consider reflected waves. In this case the time independent current distribution should be of the form of a standing wave, i.e.,

$$I(z,\omega) = I_0(\omega)\{\exp(i\omega|z|/c) - \exp[i\omega(2L-|z|)/c]\}. \tag{4.18}$$

The time dependent far-zone electric field produced by the antenna carrying a current, (4.18), is given by

$$
\begin{aligned}
E(r,t) = \frac{\theta_0\eta_0}{(2\pi r \sin\theta)} \{ & I_0(t-r/c) + I_0(t-r/c-2L/c) \\
& -I_0[t-r/c-L(1-\cos\theta)/c] \\
& -I_0[t-r/c-L(1+\cos\theta)/c]\}
\end{aligned}
\tag{4.19}
$$

where $I_0(t)$ is as expressed (4.16). Physically, the difference between (4.15) and (4.19) is that the latter indicates that an additional wavelet originating from the center of the antenna and delayed in time by $2L/c$ (the transit time across the total length of the antenna) contributes to the far-zone field. For L/c small compared to the pulse duration of the initial current wave (4.19) is given approximately by

$$E(r,t) \simeq \frac{\theta_0\eta_0}{(2\pi r)}\left(\frac{L}{c}\right)^2 [\partial^2 I(t-r/c)/\partial t^2]\sin\theta. \tag{4.20}$$

It is clear from the above that the time domain far-zone electric field of a linear antenna depends very much on the current mode excited on the antenna.

4.2.2 The Receiving Antenna

Unless the probe used to measure the electric or the magnetic field is truly independent of frequency, the received signal, represented by a load voltage or load current, will not preserve the waveform of the incident field. To investigate the response of a receiving antenna to a transient field we can again apply the method of Fourier transform to the time harmonic load voltage or current. The frequency domain equivalent circuit for a receiving antenna is known to be of the form shown in Fig. 4.3, where $V_{op} = E^i(\omega) \cdot h_r(\omega)$ is the open circuit voltage,

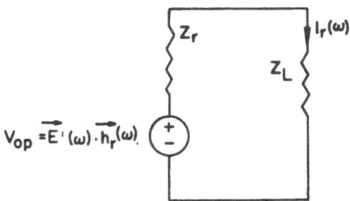

Fig. 4.3. Equivalent circuit of a receiving antenna

$E^i(\omega)$ is the incident electric field, $h_r(\omega)$ is the vector effective height of the antenna operating in the transmitting mode, Z_r is the input impedance of the receiving antenna operating in the transmitting mode, and Z_L denotes the load impedance.

Sometimes it is found to be convenient to express the open-circuit voltage in terms of the incident magnetic field, particularly for loop-type receiving antennas. In that case we define a vector effective area A such that

$$V_{op} = i\omega\mu_0 H^i(\omega)\cdot A(\omega)\,. \tag{4.21}$$

Since $H^i(\omega) = (1/\eta_0)E^i(\omega)\times r'_0$, where r'_0 denotes the unit vector pointing opposite to the direction of incidence of the incoming wave, the relation between h and A is obtained as

$$A(\omega) = (i/k)(r'_0 \times h)\,, \tag{4.22}$$

where $k = \omega/c$ is the wavenumber. We may mention that the vector effective area of an electrically small loop with axis pointed in the z-direction is given by

$$A(\omega) = \theta_0 S \sin\theta\,, \tag{4.23}$$

where S is the geometrical area of the loop. It can be shown that the vector effective height of a short dipole carrying a current distribution of the form $I(z) = I_0(1 - |z|/L)$, $|z| \leq L$ is given by

$$h(\omega) = -\theta_0 L \sin\theta\,. \tag{4.24}$$

Based on the equivalent circuit of the receiving antenna we have

$$I_r(\omega) = E^i(\omega)\cdot h_r(\omega)/(Z_r + Z_L)\,, \tag{4.25}$$

hence

$$I_r(t) = \frac{1}{2\pi}\int_{-\infty}^{\infty} E^i(\omega)\cdot h_r(\omega)(Z_r + Z_L)^{-1}\exp(-i\omega t)d\omega\,. \tag{4.26}$$

We now consider the case where the incident electric field originates from a transmitting antenna of vector effective height $h_t(\omega)$, so that

$$E^i(\omega) = i\omega\eta_0 I_t(\omega) h_t(\omega) \exp(i\omega r/c)/4\pi cr, \qquad (4.27)$$

where $I_t(\omega)$ denotes the input current to the transmitting antenna. Using (4.27) in (4.26) we obtain the following for the received current

$$I_r(t) = \frac{i\eta_0}{8\pi^2 cr} \int_{-\infty}^{\infty} \omega I_t(\omega) h_r(\omega) \cdot h_t(\omega)(Z_r + Z_L)^{-1} \exp[-i\omega(t - r/c)] d\omega. \qquad (4.28)$$

To Illustrate the significance of (4.28) we consider first the case of two idealistic linear traveling wave antennas placed parallel to each other in the broadside direction $\theta = \pi/2$. The vector effective heights of these antennas in the broadside direction are, respectively.

$$h_t(\omega) = -i2c/\omega[1 - \exp(i\omega L_1/c)] z_0 \qquad (4.29)$$

$$h_r(\omega) = -i2c/\omega[1 - \exp(i\omega L_2/c)] z_0, \qquad (4.30)$$

where L_1 and L_2 are the half-lengths of the transmitting and receiving antennas, and z_0 is the unit vector in the z-direction. Substituting (4.29), (4.30) into (4.28) and assuming that both Z_r and Z_L are purely resistive we find that the received current is given by

$$\begin{aligned} I_r(t) = \frac{\eta_0 c}{[2\pi^2 r(Z_r + Z_L)]} \int_{-\infty}^{\infty} U(t - t')\{I_t(t' - r/c) \\ - I_t[t' - (r/c) - (L_1/c)] I_t[t' - (r/c) - L_2/c)] \\ + I_t[t' - (r/c) - (L_1 + L_2)/c]\} dt'. \end{aligned} \qquad (4.31)$$

The appearance of the unit step function $U(t - t')$ in (4.31) results from the application of the convolution theorem to evaluate (4.28). The wavelets involved in (4.31) are shown schematically in Fig. 4.4. They represent waves originating from the terminals of the transmitting antenna and reaching the receiving antenna terminals through various alternative paths.

We now consider the characteristics of the voltage probes and current probes for measuring transient fields. To measure the electric field one often uses a short linear antenna. The vector effective height of a short antenna under the assumption of a linear current distribution may be obtained from (4.30). Thus the time domain response of the open circuit voltage is given by

$$V_{op}(t) = \frac{1}{2\pi} \int\limits_{-\infty}^{\infty} E^i(\omega) \cdot h(\omega) \exp(-i\omega t)d\omega = -LE^i_\theta(t)\sin\theta, \quad (4.32)$$

where $E_\theta(t)$ is the θ-component of the incident electric field. A voltage probe, therefore, would reproduce the wave form of the incident electric field provided that the frequency spectrum of the incident electric field lies mainly in the low frequency region so that the concept of a short antenna is still applicable.

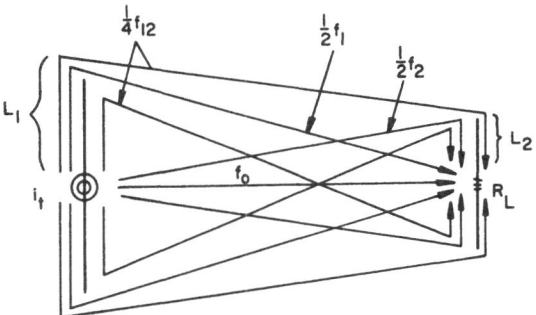

Fig. 4.4. The origin of various waves: $f_0 = f(t - r/c); f_1 = f(t - r/c - L_1/c); f_2 = f(t - r/c - L_2/c);$ $f_{12} = f[t - r/c - (L_1 + L_2)/c]$. Waves following alternative paths but suffering the same phase delay contribute to the received field. Hence the occurrence of the amplitude factors 1/2, 1/4, etc.

To investigate the response of a current probe we consider the reception of an incident magnetic field by a loop with axis pointed in the z-direction. We assume that the spectral density od the incoming signal is confined to the low frequency region so that the vector effective area of the loop can be considered, as given by (4.23), and the open circuit voltage induced in the loop is equal to $i\omega\mu_0 H^i(\omega) \cdot A(\omega)$. We assume that Z_r is approximately equal to $-i\omega L'_1$, where L'_1 is the effective inductane of the loop, and $Z_L = 0$; then (4.26) yields

$$I_r(t) = \frac{1}{2\pi} \int\limits_{-\infty}^{\infty} [i\omega\mu_0 H(\omega) \cdot A(\omega)](1/Z_r + Z_L)\exp(-i\omega t)d\omega$$

$$(4.33)$$

$$= \frac{-i\mu_0}{2\pi L'_1} \int\limits_{-\infty}^{\infty} H^i_\theta(\omega)S\sin\theta\exp(-i\omega t)d\omega - (\mu_0 S\sin\theta)H^i_\theta(t)/L'_1,$$

where S is the geometrical area of the loop and $H_\theta(t)$ is the θ-component of the incident magnetic field. Under the conditions which we have assumed, the received current would reproduce the waveform of the incident magnetic (or electric) field. The few examples given here, although idealistic, do show the basic processes involved in the transmission and reception of transient signals.

4.2.3 Directivity

For a monochromatic time harmonic field the directivity of an antenna is a well defined and useful parameter in antenna theory. In terms of the vector effective height defined previously, we have the directivity D given by

$$D = |h(\omega)|^2 \left(\frac{1}{4\pi} \int_0^{4\pi} |h(\omega)|^2 \, d\Omega \right)^{-1}, \qquad (4.34)$$

where $d\Omega$ denotes the solid angle with apex at the origin of the coordinate system $(r = 0)$. For transient signals, the ordinary concept of directivity is no longer applicable. In fact, the interference phenomenon which is responsible for producing the directivity of an antenna for a harmonically oscillating field is not present in video pulsed signals. To define the directivity of an antenna emitting transient signals, it seems reasonable that we use the energy content of the signal as the basic measure. One definition which we propose here is

$$D = \left(\int_{-\infty}^{\infty} |E(t)|^2 \, dt \right) \left(\frac{1}{4\pi} \int_0^{4\pi} \int_{-\infty}^{\infty} |E(t)|^2 \, dt \, d\Omega \right)^{-1}. \qquad (4.35)$$

The numerator in (4.35) may be identified with the total radiated energy in a certain direction and the denominator with the total energy radiated in all directions. As a result of Parseval's theorem [4.31], (4.35) can be written as

$$D = \left(\int_{-\infty}^{\infty} |E(\omega)|^2 \, d\omega \right) \left(\frac{1}{4\pi} \int_0^{4\pi} \int_{-\infty}^{\infty} |E(\omega)|^2 \, d\omega \, d\Omega \right)^{-1}. \qquad (4.36)$$

For the type of field described by (4.17) and (4.20), the directivity so calculated is the same as that of a short dipole for monochromatic radiation. For transient radiation, the pulse duration, therefore, plays an important role in determining the directivity of an antenna.

4.3 Fourier Transform Method of Solution

The classical method of solving the problem of transient radiation from linear antennas consists in analyzing the pertinent antenna characteristics by standard frequency domain techniques with a subsequent Fourier transformation to obtain the corresponding time domain solution. A complete determination of the time response for any input signal requires the knowledge of the frequency response up to the highest frequency for which the input signal has significant energy. Obtaining the monochromatic input and radiation characteristics of an antenna is essentially a problem of solving Maxwell's equations subject to the boundary conditions imposed by the antenna and the source. Even for a simple linear antenna, analytical determination of the frequency domain characteristics over the entire band of frequencies of interest is extremely difficult, if not impossible. It is for this reason that numerical methods are often used to obtain solutions. The development of Fast Fourier Transform (FFT) techniques [4.32] and sufficient computing capabilities in modern machines have made it possible to obtain solutions to practical problems. Although the final goal of the analysis is to obtain time domain results, the frequency domain results, in particular the transfer function of the antenna and the spectral density of the radiated wave form, yield much interesting information which other methods (for example, direct time domain analysis) may not yield.

4.3.1 Basic Relations

Let a linear antenna of length $2L$ be aligned along the z-axis of a rectangular coordinate system with the origin located at the center of the antenna as shown in Fig. 4.1, and let it be excited by a unit slice generator located at the origin. (A slice generator of strength V_0 is defined to be such that when connected across the delta-gap at the center of the antenna, it maintains a field $E_z = -V_0\delta(z)$ in the gap, where $\delta(z)$ is the Dirac delta function.) The far electric field produced by such an antenna is directed entirely in the θ-direction. In the time harmonic case the far electric field is given by

$$F = F'_\theta(r,\theta,\omega)\exp(-i\omega t), \tag{4.37}$$

where

$$F'_\theta(r,\theta,\omega) = \frac{-i\omega\eta_0\sin\theta}{4\pi c r}\exp(i\omega r/c)\int_{-L}^{L} I(z',\omega)\exp(-i\omega z' c^{-1}\cos\theta)dz', \tag{4.38}$$

$I(z',\omega)$ is the current distribution on the antenna due to the harmonically time dependent unit slice generator, and the other parameters are defined as before.

In (4.38), $F_\theta'(r,\theta,\omega)$ is the time independent far electric field produced by the antenna. For convenience we define the antenna transfer function in the following manner:

$$F_\theta(\theta,\omega)=rF_\theta'(r,\theta,\omega)\exp(-i\omega r/c)$$

$$=\left(\frac{-i\omega\eta_0\sin\theta}{4\pi c}\right)\int_{-L}^{L}I(z',\omega)\exp(-i\omega z'c^{-1}\cos\theta)dz'.$$

(4.39)

Notice that in the transfer function defined by (4.39), the dependence of the far-field amplitude on r as well as the phase shift suffered by the field in traveling from the antenna to the far-field point are both suppressed. Consequently, all the field quantities, in our subsequent discussion, will be independent of r.

Let the slice generator have arbitrary time dependence such that the input signal voltage in time is represented by $V(t)$. It is assumed that $V(t)$ is Fourier transformable, i.e.,

$$V(\omega)=\int_{-\infty}^{\infty}V(t)\exp(i\omega t)dt,$$

(4.40)

$$V(t)=\frac{1}{2\pi}\int_{-\infty}^{\infty}V(\omega)\exp(-i\omega t)d\omega.$$

(4.41)

If the input signal is Gaussian in time, we have

$$V_g(t)=\exp(-t^2/2\sigma^2),$$

(4.42)

where σ is proportional to the width of the input pulse. The Fourier transform of the input pulse given by (4.42) is

$$V_g(\omega)=(2\pi)^{-1/2}\sigma\exp(-\omega^2\sigma^2/2).$$

(4.43)

For a rectangular input signal of width T we have

$$V_r(t)=\begin{cases}1/T & \text{for } 0\le t\le T\\ 0 & \text{otherwise}.\end{cases}$$

(4.44)

The frequency spectrum of the signal given by (4.44) is

$$V_r(\omega) = [(\sin \omega T/2)/(\omega T/2)] \exp(i\omega T/2) \qquad (4.45)$$

If the input signal is a delta function in time then its frequency spectrum $V_\delta(\omega) = 1$.

After making use of the linearity of the system along with the super-position theorem and the inverse Fourier transform, one may show that the time dependent far field produced by the antenna excited by a slice generator signal $V(t)$ si given by

$$
\begin{aligned}
E_\theta(\theta, t) &= \frac{1}{2\pi} \int_{-\infty}^{\infty} F_\theta(\theta, \omega) V(\omega) \exp(-i\omega t) d\omega \\
&= \frac{1}{2\pi} \int_{-\infty}^{\infty} E_\theta(\theta, \omega) \exp(-i\omega t) d\omega .
\end{aligned}
\qquad (4.46)
$$

In (4.46), the quantity $E_\theta(\theta, \omega) = F_\theta(\theta, \omega) V(\omega)$ may be looked upon as the spectral density of the far-field waveform. Notice that by definition, $E_\theta(\theta, t)$ and $E_\theta(\theta, \omega)$ are related to each other by the usual Fourier transform relations.

As can be seen from the above, the first step in the analysis is the determination of the current distribution $I(z', \omega)$ on the antenna when it is excited by a harmonically time dependent unit slice generator. Let us assume for generality that the antenna is loaded with distributed resistance such that its internal impedance may be expressed as $Z^i(z')$ ohms/meter for $0 \le |z'| \le L$. Under this condition [4.33] it can be shown that the current distribution $I(z', \omega)$ on the antenna satisfies the following modified form of Hallen's integral equation:

$$
\int_{-L}^{L} I(z', \omega)(4\pi)^{-1}[(z-z')^2 + a^2]^{-1/2} \exp\{i(\omega/c)[(z-z')^2 + a^2]^{1/2}\} dz'
\qquad (4.47)
$$
$$
= B\cos kz + i(2\eta_0)^{-1}\sin k|z| - i\eta_0^{-1} \int_0^z I(\xi, \omega) Z^i(\xi) \sin[k(z-\xi)] d\xi
$$

where $k = \omega/c$ is the propagation constant, B is a constant to be determined from the end condition $I(\pm L, \omega) = 0$, and a is the radius of the antenna element. The usual thin wire approximations have been made in deriving (4.47)[1].

Successful application of the Fourier transform method implies the knowledge of the transfer function of the antenna as a function of frequency. This requires the determination of the time harmonic current

---

[1] Thin wire approximations imply that the current on the antenna is constrained along the z-axis and also that $L \gg a$, $ka \ll 1$.

distribution on the antenna at all frequencies of interest. Only in very exceptional cases is it possible to obtain the solutions analytically. For this reason, numerical techniques are used quite often to obtain the desired current distribution. In the next Subsection we give a brief discussion of the numerical technique that may be used for such purposes and for obtaining the final time dependent radiated wave forms.

4.3.2 Outline of the Numerical Method

Standard numerical techniques [4.34, 35] are used to solve the integral equation (4.47) for the time harmonic current distribution $I(z, \omega)$. For this purpose the integral equation is reduced by moment methods to the following set of N simultaneous algebraic equations

$$\sum_{n=1}^{N} \int_{\Delta z_n} I(z', \omega) G(z_j, z') dz'$$

$$= B \cos k z_j + (i/2\eta_0) \sin k |z_j| - (i/\eta_0) \sum_{n}' I(z', \omega) Z^i(z') \sin k(z_j - z') dz',$$

$$j = 1, 2, \ldots, N, \tag{4.48}$$

where

$$G(z_j, z') = (4\pi)^{-1} [(z_j - z)^2 + a^2]^{-1/2} \exp\{ik[(z_j - z)^2 + a^2]^{1/2}\}, \tag{4.49}$$

$$\sum_{n}' = \begin{cases} \sum_{n=j}^{N/2}, & \text{for } j \le N/2 \\ \sum_{n=(N/2)+1}^{j} & \text{for } j > N/2 \end{cases} \tag{4.50}$$

and it is assumed that N is an even number and $z' \in \Delta z_n$. Equation (4.49) implies that the antenna of length $2L$ is divided into N sections; the number of sections increases from 1 to N along the antenna from $-L$ to L.

It is now necessary to make an appropriate approximation to the current distribution $I(z')$ (for simplicity we use the notation $I(z')$ for $I(z', \omega)$) in each section Δz_n. Various choices are available for this approximation [4.34,35]. We make the following quadratic approximation to the unknown current in each section

$$I(z') = \begin{cases} A_n + B_n(z' - z_n) + C_n(z' - z_n)^2, & \text{for } z' \in \Delta z_n \\ 0, & \text{otherwise} \end{cases} \tag{4.51}$$

where A_n, B_n, C_n are three unknown constants. These constants are determined by requiring that the continuation of the expression given by (4.51) into the centers of the adjacent sections give the appropriate values of the currents there. After evaluating the constants it can be shown [4.33] that the current in each section is expressed by the recurrence relation

$$I(z') = I_{n-1} X_n(z') + I_n Y_n(z') + I_{n+1} Z_n(z'), \qquad z' \in \Delta z_n \qquad (4.52)$$

where

$$X_n(z') = -(z' - z_n)(2\Delta z)^{-1} + (z' - z_n)(\Delta z)^{-2}/2, \quad z' \in \Delta z_n \qquad (4.53)$$

$$Y_n(z') = 1 - (z' - z_n)^2(\Delta z)^{-2}, \qquad\qquad\qquad z' \in \Delta z_n \qquad (4.54)$$

$$Z_n(z') = (z' - z_n)(2\Delta z)^{-1} + (z' - z_n)^2(\Delta z)^{-2}/2, \quad z' \in \Delta z_n \qquad (4.55)$$

$$I(z_n) = I_n \quad \text{etc.}, \qquad\qquad\qquad\qquad\qquad\qquad (4.56)$$

Δz is the length of each section Δz_n, and $n \neq 1, N$. For the two end sections Δz_1 and Δz_N, the currents are given by [4.33]

$$I(z') = I_1 Y_1'(z') + I_2 Z_1'(z') \qquad \text{for} \quad z' \in \Delta z_1 \qquad (4.57)$$

$$I(z') = I_N Y_N'(z') + I_{N-1} X_N'(z') \quad \text{for} \quad z' \in \Delta z_N \qquad (4.58)$$

where

$$Y_1'(z') = 1 + (z' - z_1)^2(\Delta z)^{-1} - 2(z' - z_1)^2(\Delta z)^{-2}, \qquad \text{for} \quad z' \in \Delta z_1$$
$$(4.59)$$

$$Z_1'(z') = (z' - z_1)(3\Delta z)^{-1} + 2(z' - z_1)^2(\Delta z)^{-2}/3, \qquad \text{for} \quad z' \in \Delta z_1$$
$$(4.60)$$

$$Y_N'(z') = 1 - (z' - z_N)(\Delta z)^{-1} - 2(z' - z_N)^2(\Delta z)^{-2}, \qquad \text{for} \quad z' \in \Delta z_N$$
$$(4.61)$$

$$X_N'(z') = -(z' - z_N)(3\Delta z)^{-1} + 2(z' - z_N)^2(\Delta z)^{-1}/3, \quad \text{for} \quad z' \in \Delta z_N$$
$$(4.62)$$

After substituting (4.51)–(4.62) into (4.48) we obtain a set of N simultaneous equations involving the N unknown current coefficients $I_1, I_2, \ldots, I_N$. The extra unknown coefficient B in (4.48) is now determined by applying the end condition $I(L,\omega)=0$. By using Taylor series expansion for the currents at the centers of the last four sections and retaining the first four terms in each series we obtain the following additional equation:

$$-5I_{N-3}+21I_{N-2}-35I_{N-1}+35I_N=0. \qquad (4.63)$$

We now have $N+1$ equations for the $N+1$ unknowns and the system of equations is solved by conventional matrix methods for $I_1, I_2, ..., I_N$ and B.

The transfer function $F_\theta(\theta,\omega)$ given by (4.39) can now be evaluated numerically with the help of the relation

$$F_\theta(\theta,\omega) = -(i\omega\eta_0 \sin\theta/4\pi c) \sum_{n=1}^{N} I_n \Delta z_n \exp\left[-(i\omega z_n \cos\theta)/c\right] \quad (4.64)$$

where we have used the notation $I(z_n,\omega)=I_n$, and $z'=z_n$ is the coordinate of the center of the section Δz_n.

The spectral density $E_\theta(\theta,\omega)$ of the radiated wave forms is obtained by using the relation $E_\theta(\theta,\omega)=F_\theta(\theta,\omega)V(\omega)$. Finally the time dependent radiated wave form $E_\theta(\theta,t)$ is obtained by carrying out the integration given by (4.46) with the help of the fast Fourier inversion technique. Details of FFT techniques may be found in [4.32].

During the numerical computation as noted above, it is found convenient to divide the antenna length into N equal length sections. The frequency domain calculations may be truncated at the highest frequency f_0 such that $\Delta z = 2L/N \simeq \lambda_0/6$, where λ_0 is the free space wavelength corresponding to f_0. In Section 4.5 we shall discuss some results obtained by the application of the method described here.

4.4 Analytical Solution for a Reflectionless Linear Antenna

As an application of the Fourier transform method discussed in Section 4.3 we consider the transient wave forms radiated by linear antennas loaded non-uniformly and continuously with resistance. The antennas considered are thin cylinders excited symmetrically at the centers by slice generators having a Gaussian pulse type of time dependence. Emphasis is given to a specific type of resistive loading which makes the antenna non-reflecting [4.36] and for which an analytical expression for the time harmonic current distribution is known. The choice of Gaussian input pulse and the specific type of loading makes it possible to obtain closed form theoretical expressions for the wave forms radiated by the antenna under consideration. The discussion given here follows closely that in [4.18].

4.4.1 Loading of the Antenna and Current Distribution

The antenna is assumed to be resistively loaded such that its internal impedance may be expressed as

$$Z^i(z') = \alpha_0/(1 - |z'|/L) \, [\Omega/\text{m}], \quad -L \leq z' \leq L, \tag{4.65}$$

where $\alpha_0 = Z^i(0)/L \, [\Omega/\text{m}]$ is referred to as the loading parameter. Equation (4.65) implies that the internal resistance of the antenna element increases continuously from the value α_0 at the input end to infinity at the antenna end points. As discussed in [4.37], the above type of loading may be realized in practice by distributed loading. The rationale behind this particular type of resistive loading is as follows. It is known [4.36] that for a linear antenna loaded according to (4.65) there exists a critical value of α_0, depending on the thickness parameter $\Omega = 2\ln(2L/a)$ of the antenna, for which the loaded antenna excited by a harmonic slice generator sustains a single wave of current traveling from the generator toward both ends of the antenna. For this reason the antenna loaded according to (4.65) and with the critical value of α_0 is sometimes referred to as the reflectionless antenna. For example, if $\alpha_0 L = Z^i(0) = (\eta_0/\pi)\ln(2L/a)$, the current distribution on the harmonically excited antenna is given by

$$I(z', \omega) = (1/\alpha_0 L)[-i\omega\tau/(1 - i\omega\tau)](1 - |z'|/L)\exp[i(\omega/c)|z'|] \tag{4.66}$$

where $\tau = L/c$ is the transit time of the signal on the antenna. For α_0 less than the critical value the antenna stops being reflectionless and sustains standing wave current distribution. From theoretical considerations it has been found [4.38] that for α_0 larger than the critical α_0, the antenna sustains a progressive wave of current whose distribution may be expressed as a hypergeometric function.

For a Gaussian input signal of the form given by (4.42) and (4.43), the time dependent current distribution on the reflectionless antenna is obtained from

$$I(z,t) = \frac{1}{2\pi} \int_{-\infty}^{\infty} I(z,\omega) V_g(\omega)\exp(-i\omega t)d\omega, \tag{4.67}$$

where $V_g(\omega)$ is given by (4.43). The evaluation of the integral in (4.67) is discussed in [4.18]. The final expression for the time dependent current distribution is

$$I(z,t) = (2\pi)^{-1/2}(1/\alpha_0 L)(1 - |z|/L)[2^{-1/2}\exp[-(t-|z|/c)^2/2\sigma^2]$$
$$- (\sigma/\tau)\exp(\sigma^2/2\tau^2)\{1 + \mathrm{erf}[2^{-1/2}\sigma^{-1}(t-z/c)]\}$$
$$\times \exp[-\tau^{-1}(t-z/c)]] \tag{4.68}$$

where

$$\mathrm{erf}(x) = 2\pi^{-1/2} \int_0^x \exp(-t^2) \tag{4.69}$$

4.4.2 Spectral Density of the Wave Form

Using (4.39), (4.46) and (4.43) the spectral density of the wave forms radiated by the antenna may be obtained. It can be shown [4.18] that the spectral density is given by the following expression

$$E_\theta(\theta,\omega) = (2\pi)^{-1/2}(\eta_0\sigma)(2\alpha_0 L)^{-1}\sin\theta[(i\omega\tau)^2/(1 - i\omega\tau)]\exp(-\sigma^2\omega^2/2)$$
$$\times [(i\omega\tau v)^{-1}\{[\exp(i\omega\tau v) - 1](i\omega\tau v)^{-1} - 1\} \tag{4.70}$$
$$+ (i\omega\tau v')^{-1}\{[\exp(i\omega\tau v') - 1](i\omega\tau v')^{-1} - 1\}],$$

$$v = 1 - \cos\theta, \quad v' = 1 + \cos\theta, \quad \tau = L/c. \tag{4.71}$$

All the important frequency domain results pertinent to the antenna may be obtained from (4.70). It is interesting to investigate analytically the high and low frequency behavior of the spectral density. It can be shown that in the low and high frequency limits (4.70) behaves asymptotically as

$$E_\theta(\theta,\omega) \simeq (2\pi)^{-1/2}\eta_0\sigma(2\alpha_0 L)^{-1}(-i\omega\tau^{-1})\exp(-\omega^2\sigma^2/2)\sin\theta \tag{4.72}$$
$$\text{for} \quad \omega\tau \ll 1$$

$$E_\theta(\theta,\omega) \simeq (2\pi)^{-1/2}\eta_0\sigma(\alpha_0 L)^{-1}\exp(-\omega^2\sigma^2/2)/\sin\theta \tag{4.73}$$
$$\text{for} \quad \omega\tau \gg 1, \; \theta \neq 0.$$

Equations (4.72) and (4.73) indicate that for all $\theta \neq 0$, $E_\theta(\theta,\omega) \to 0$ as $\omega \to 0$ and $\omega \to \infty$. Further study of (4.70) in the region $\theta \simeq 0$ indicates that $E_\theta(\theta,\omega)\to 0$ for $\theta=0$ for all ω. Thus it may be concluded that for all values of θ, the spectral density tends to zero in both the low and high frequency limits. This is consistent with the physical aspects of the radiation mechanism from the antenna at low frequencies and with the band-limited nature of the input wave form.

4.4.3 Time-Dependent Radiated Wave Form

Introducing (4.70) into (4.46) yields the time dependent far-field wave form radiated by the antenna in the following form

$$E_\theta(\theta,t)=(2\pi)^{-1/2}(\eta_0\sigma)(2\alpha_0 L)^{-1}\sin\theta[A(v,t)+A(v',t)]\qquad(4.74)$$

where

$$A(v,t)=\frac{1}{2\pi}\int_{-\infty}^{\infty}\exp(\sigma^2\omega^2/2)(i\omega\tau)^2(1-i\omega\tau)^{-1}$$
$$\cdot\{-(i\omega\tau v)^{-1}+[\exp(i\omega\tau v)-1](i\omega\tau)^{-2}v^{-2}\{\exp(-i\omega t)d\omega.$$
$$(4.75)$$

The integral in (4.75) may be carried out explicitly, the details of which may be found in [4.18]. The final expression for the far-field wave form is given by

$$E_\theta(\theta,t)=(2\pi)^{-1/2}\eta_0(4\alpha_0 L)^{-1}\sin\theta(2^{-1/2}(v^{-1}+v'^{-1})\exp(-t^2/2\sigma^2)$$
$$-(\sigma/\tau)\exp(-t/\tau)\exp(\sigma^2/2\tau^2)\{1+\mathrm{erf}[2^{-1/2}(t/2-\sigma/\tau)]\}$$
$$\cdot[(v+1)/v^2+(v'+1)/v'^2]+(\sigma/\tau)v^{-2}\exp(\sigma^2/2\tau^2)$$
$$\cdot\{1+\mathrm{erf}(2^{-1/2})[(t-v\tau)/\sigma-\sigma/\tau]\}\{\exp[-(t-v\tau)/\tau]\}\quad(4.76)$$
$$+(\sigma/\tau)v'^{-2}\exp(\sigma^2/2\tau^2)$$
$$\cdot\{1+\mathrm{erf}(2^{-1/2})[(t-v'\tau)/\sigma-\sigma/\tau]\}\exp[-(t-v'\tau)/\tau]).$$

In the broadside direction, $\theta=\pi/2$, $v=v'=1$ and we obtain from (4.76)

$$E_\theta(\pi/2,t)=(2\pi)^{-1/2}\eta_0(2\alpha_0 L)^{-1}$$
$$\times[2^{-1/2}\exp(-t^2/2\sigma^2)-(2\sigma/\tau)\exp(\sigma^2/2\tau^2)$$
$$\times\{1+\mathrm{erf}[2^{-1/2}(t/\sigma-\sigma/\tau)]\exp(-t/\tau)+(\sigma/\tau)\exp(\sigma^2/2\tau^2)$$
$$\times(1+\mathrm{erf}\{2^{-1/2}[(t-\tau)/\sigma-\sigma/\tau]\})\exp[-(t-\tau)/\tau]].$$
$$(4.77)$$

For both large positive and negative time, i.e., $|t| \gg \tau$ and $\tau > \sigma$ it can be shown from (4.77) that

$$E_\theta(\pi/2, t) \cong (2\pi)^{-1/2} \eta_0 (2\alpha_0 L)^{-1} \{2^{-1/2} \exp(-t^2/2\sigma^2) \tag{4.78}$$
$$+ (\sigma/\tau) \exp(\sigma^2/2\tau^2) [1 + \mathrm{erf}(t2^{-1}\sigma^{-1})](e-2)\exp(-t/\tau)\},$$
$$\text{for} \quad |t| \gg \tau, \, \tau > \sigma \, .$$

The general expression (4.76) indicates that the radiated wave form is different from Gaussian and also that it depends strongly on the parameter (σ/τ). The time asymmetry in the radiated wave form (i.e., $E_\theta(\theta, t) \neq E_\theta(\theta, -t)$) follows from the fact that in the present case the antenna transfer function is not a real and even function of frequency. Physically the parameter σ/τ determines the effective illumination of the antenna and hence it is anticipated that the radiated pulse shape will strongly depend on the ratio of the input pulse width to the transit time on the antenna. It is interesting to observe that from (4.76) we obtain the following in the limit $\tau \to \infty$

$$E_\theta(\theta, t) \simeq (2\pi)^{-1/2} \eta_0 (4\alpha_0 L)^{-1} (\sin\theta)(1/v + 1/v') \exp(-t^2/2\sigma^2) \, ,$$
$$\text{for} \quad \tau \to \infty \, . \tag{4.79}$$

Thus, the wave form radiated by the infinitely long antenna is also Gaussian. This result should be compared with the radiated wave form from an infinitely long unloaded linear antenna excited by a Gaussian signal [4.19]. Although the radiated wave form is found to be almost Gaussian in the unloaded case, the time symmetry of the wave form is slightly lost due to the nature of the antenna transfer function mentioned earlier.

Figures 4.5 and 4.6 show the radiated wave forms in the broadside direction as functions of t/τ for $\sigma = 1.0$ ns and $\sigma = 0.471$ ns, respectively, for two different values of the loading parameter α_0. The antenna length is 2 meters so that the transit time $\tau = 3.33$ ns. The corresponding values of $E_\theta(\pi/2, t)$ obtained by numerically solving the integral equation for time-harmonic current and using the result in (4.64) are also shown for comparison. The results indicate that the agreement between the analytical and numerical solutions is satisfactory over the initial regions of the radiated wave forms. Near the negative peak the two results differ appreciably. This disagreement is attributed to the idealized nature of the current distribution assumed in the analytical solution. However, the overall shapes of the radiated wave forms are predicted fairly well by the analytical solution.

Observe that the negative peaks in the radiated wave forms occur at approximately $t/\tau \simeq 1$ unit after the initial peaks in Fig. 4.5, but at approximately $t/\tau \simeq 0.5$ unit for Fig. 4.6. The predominantly double peak nature of the radiated wave forms indicates that the antenna

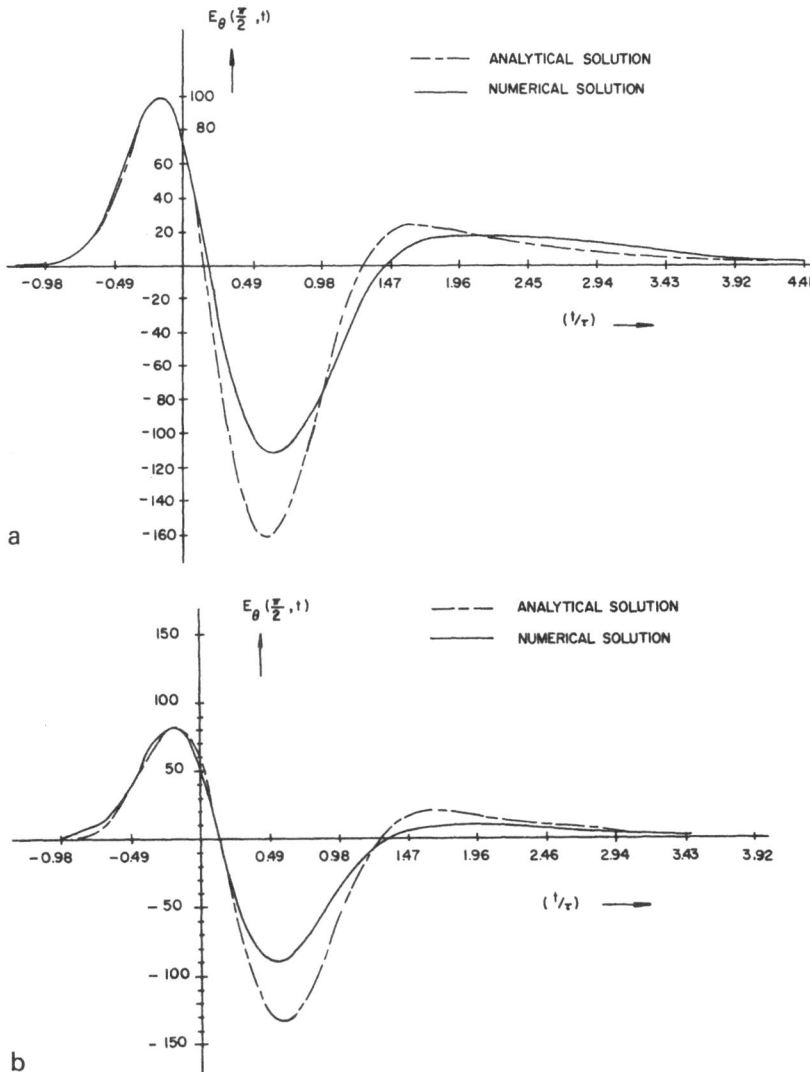

Fig. 4.5a and b. Far-field wave form in the broadside direction (after SENGUPTA and LIU [4.18]). a) $\sigma = 1$ ns, $\tau = 3.3$ ns, $\alpha_0 = 292$, $L/a = 156$. b) $\sigma = 1$ ns, $\tau = 3.33$ ns, $\alpha_0 = 464$, $L/a = 156$

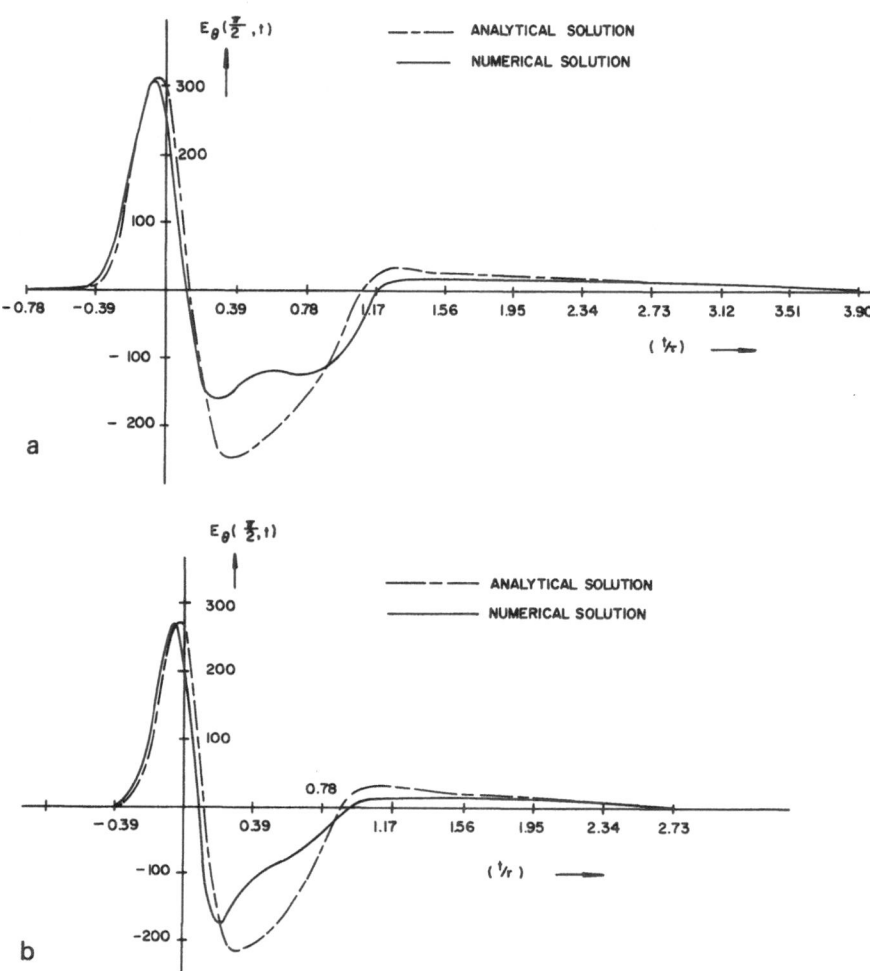

Fig. 4.6a and b. Far-field wave form in the broadside direction (after Sengupta and Liu [4.18]). a) $\sigma = 0.471$ ns, $\tau = 3.33$ ns, $\alpha_0 = 292$, $L/a = 156$. b) $\sigma = 0.471$ ns, $\tau = 3.33$ ns, $\alpha_0 = 464$, $L/a = 156$

differentiates the input Gaussian voltage wave form. In a differentiated Gaussian wave form the time separation between the two peaks is directly proportional to σ. This explains the smaller distance between the two peaks exhibited in Fig. 4.6. The radiated wave forms shown clearly indicate the reflectionless nature of the antenna. Wave forms in other directions and more detailed discussion about the antenna may be found in [4.18]. The results given here may be used to obtain some useful information about the transient behavior of resistively loaded linear antennas, and they also may provide check cases for more general numerical methods now in use.

4.5 Numerical Solution for a Reflectionless Linear Antenna

In this section we consider the numerical solution based on the Fourier transform method of the transient radiation from the nonreflectively loaded linear antenna considered in Section 4.4. The excitation of the antenna is assumed to be a rectangular pulse of the form (4.44) and (4.45). For sufficiently large T (usually T greater than several transit times τ for the present case) the radiated wave forms may be identified with those radiated by the same antenna when excited by a unit step voltage in time. For convenience of numerical computation the rectangular pulse type of excitation has been used here to obtain the step voltage response from the antenna. The theoretical basis of the numerical method is as discussed in Subsection 4.3.2. In the following sections we give a brief discussion of some of the selected results. More detailed results and discussion of the same problem may be found in [4.22] and [4.33].

4.5.1 Time-Independent Current Distribution

It is instructive to study how the loading affects the current distribution on the harmonically excited antenna. The results shown here are obtained by numerically solving (4.49). Figures 4.7a and 4.7b show the amplitude and phase distribution, respectively, of the time independent current on the linear antenna of length $2L = 5\lambda$, $L/a \simeq 156$ and $\alpha_0 \simeq 318$. Under these conditions the antenna approximately satisfied the non-reflecting criterion [4.36, 37]. The approximate theories predict the existence of a pure traveling wave of current with linearly decaying amplitude distribution on the antenna. The results shown in Fig. 4.7a indicate that the amplitude distribution of current may be considered to be approximately linear except near the feed region where it is significantly different. The linear variation of phase in Fig. 4.7b indicates the existence of a pure traveling wave of current in the antenna.

Figures 4.8a and 4.8b show the amplitude and the phase distribution of the current on the harmonically excited antenna of length $2L = 3\lambda$, $L/a = 156$ and for different values of the loading parameter α_0. For $\alpha_0 < 318$, the reflection effects on the current distribution become appreci-

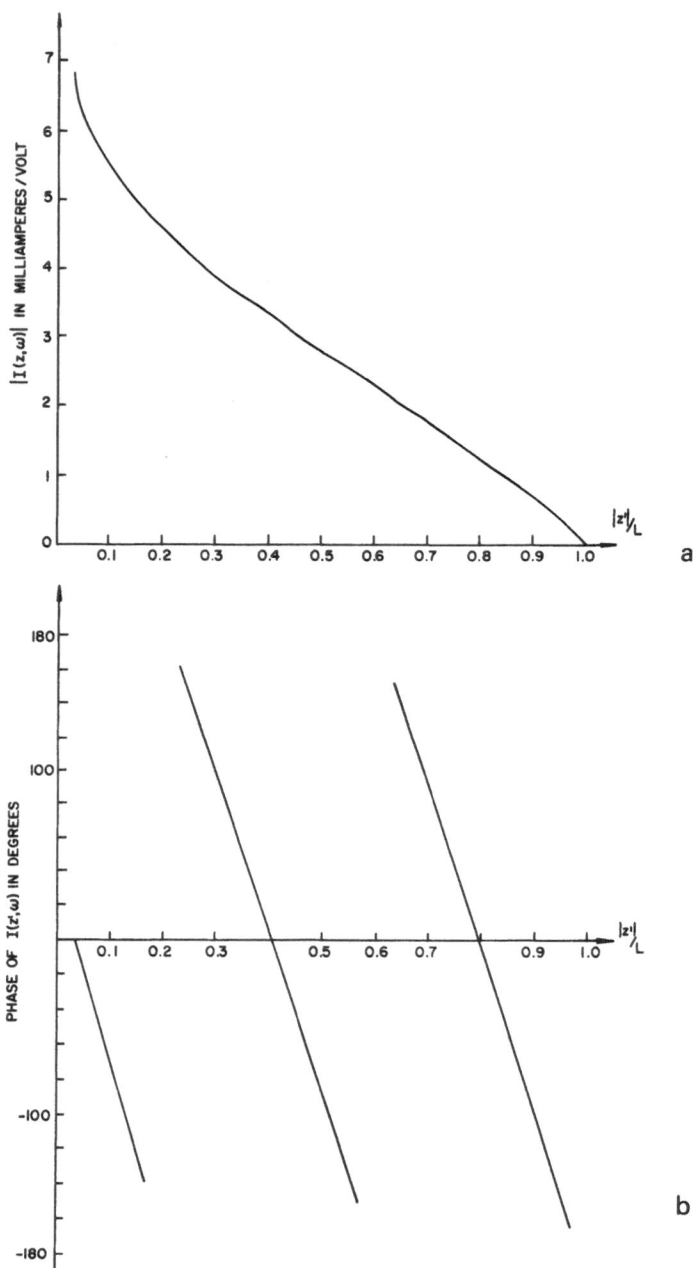

Fig. 4.7a and b. Time harmonic current distribution along a nonreflecting loaded antenna (after Lɪᴜ and Sᴇɴɢᴜᴘᴛᴀ [4.22]). $L/a = 156$, $\omega L/c = 5\pi$, $\alpha_0 = 318$. a) Amplitude. b) Phase

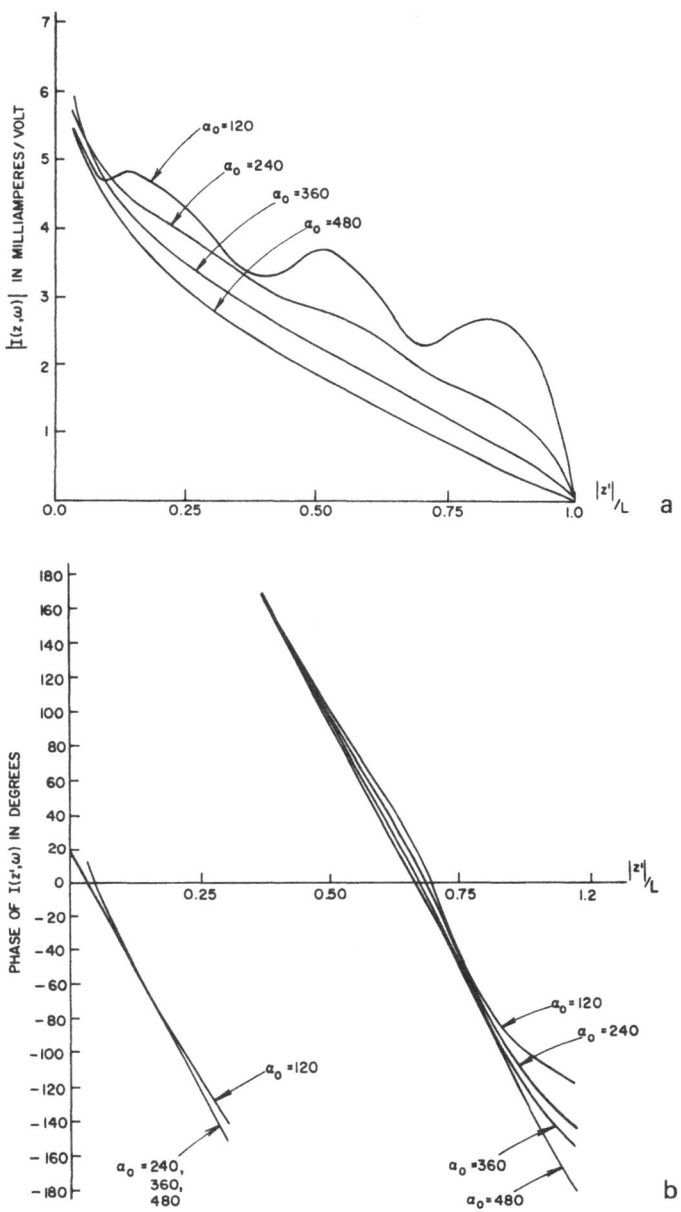

Fig. 4.8a and b. Time harmonic current distribution along a nonreflectively loaded antenna as a function of the loading parameter (after LIU and SENGUPTA [4.22]). $L/a = 156$, $\omega L/c = 3\pi$. a) Amplitude. b) Phase

able; the results for $\alpha_0 = 120$ clearly indicate that the antenna sustains a standing wave type of current distribution. For $\alpha_0 > 318$, the amplitude distribution of the current resembles an exponentially decaying function and the phase distribution, although not linear, is progressive along the length of the antenna. In this sense it is proper to say that for α_0 larger than the critical value of the loading parameter (i.e., a larger than 318 in this case) the antenna maintains its non-reflecting properties, but it supports a progressive wave of current. It should be noted that the critical value of the loading parameter α_0 is different for an antenna with a different value of L/a.

For $L/a = 156$, $\alpha_0 = 600$, it has been found [4.33] that the antenna retains its non-reflecting characteristics up to $\omega L/c = 30$ where the present computation procedure was truncated.

4.5.2 Transfer Function of the Antenna $F_\theta(\theta, \omega)$

The transfer functions of the antenna are obtained numerically from (4.64). Figure 4.9 gives the variations of the magnitude and phase of $F_\theta(\theta, \omega)$ in the broadside direction $(\theta = \pi/2)$ of the antenna of length

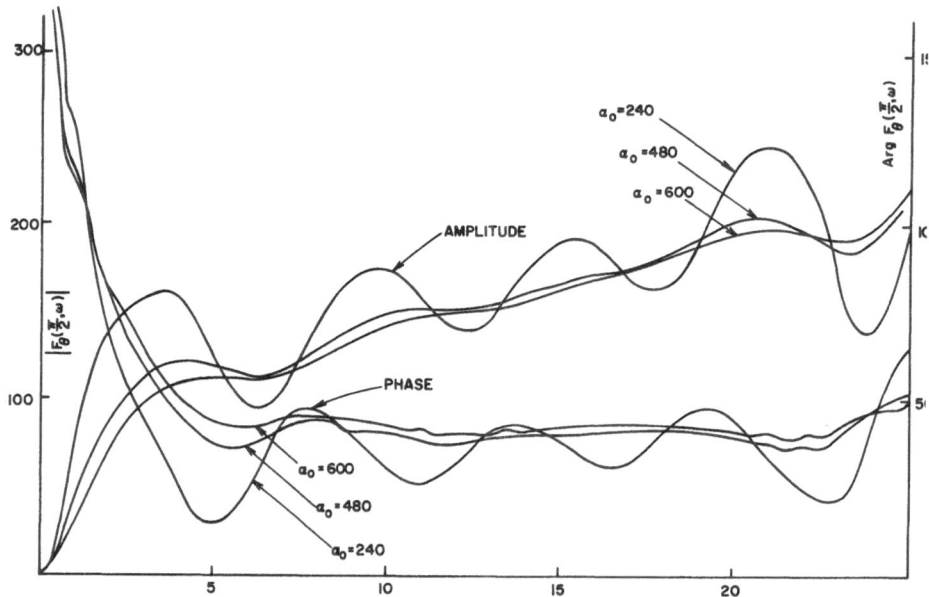

Fig. 4.9. Amplitude and phase of $F_\theta(\pi/2, \omega)$ as functions of normalized frequency with α_0 as the parameter. $L/a = 156$ (after Liu and Sengupta [4.22])

$2L$ and $L/a=156$, and for three values of the loading parameter. In all the curves, $|F_\theta(\theta,\omega)|$ approaches zero as ω approaches zero, which is consistent with the fact that there is no radiation at zero frequency. For higher frequencies the mean value of $|F_\theta(\theta,\omega)|$ tends to increase with an increase of frequency. In general, the mean value of the phase of $F_\theta(\theta,\omega)$ appears to decrease rapidly for small values of $\omega L/c$ and then assumes a constant value.

In each case, for $\alpha_0=240$, both the amplitude and phase of $F_\theta(\theta,\omega)$ oscillate with $\omega L/c$. With increase of α_0 these oscillations are smoothed out. It is interesting to observe that the oscillations in the amplitude and phase of $F_\theta(\theta,\omega)$ appear to be smoothed out for $\alpha_0\geq$ the critical loading parameter. Further details about the transfer function in other directions and for different values of L/a may be found in [4.22, 33]. In general, the increase of the parameter L/a tends to increase the amount of oscillation in the curves. For a given L/a and α_0 the oscillatory nature of the transfer function appears to be dominant in the broadside direction.

4.5.3 Spectral Density of the Radiated Wave Form

The excitation function is assumed to be of the form (4.44) with $T=20.9$ ns; the frequency spectrum is given by (4.45). As mentioned earlier for this value of the input pulse width, which is larger than τ ($=3.33$ ns), the results can approximate the step voltage excitation case.

For three given values of α_0 and $L/a=156$, Figs. 4.10a and 4.10b show the envelope of $|E_\theta(\theta,\omega)|$ vs $\omega L/c$ for two values of the observation angle θ. Detailed results for different values of L/a and various observation angles may be found in [4.22] and [4.33]. It is found that in general the envelope of $|F_\theta(\theta,\omega)|$ vs $\omega L/c$ has a dominant peak at the low frequency end for all values of θ and α_0. After the initial peak, $|E_\theta(\theta,\omega)|$ decays, at first rapidly and then slowly with increasing $\omega L/c$. Depending on the values of α_0 and θ there also appear some minor peaks for large values of $\omega L/c$. For the purpose of discussing the behavior of $|E_\theta(\theta,\omega)|$, let us define the frequency regimes that are larger and smaller than the frequency where the initial peak appears as the high and low frequency regimes. Critical study of Fig. 4.10 and the detailed results in [4.22, 33] reveals the following observations:

a) As θ decreases away from the broadside direction $\theta=90°$, the low frequency content of $|E_\theta(\theta,\omega)|$ decreases and the high frequency content decreases. Also for a given loading, as θ decreases, the amplitude of the initial peak decreases. For a given α_0, the position of the initial

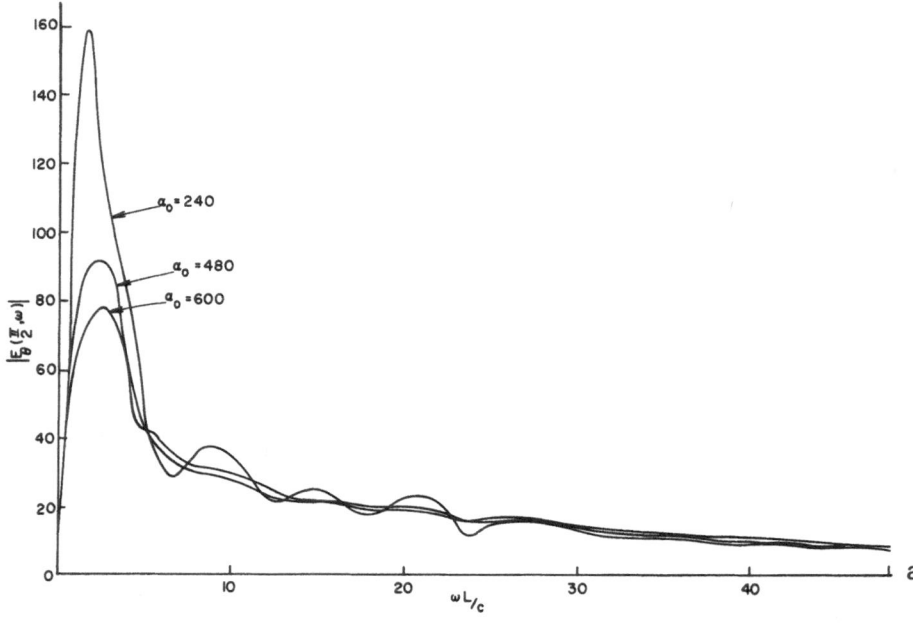

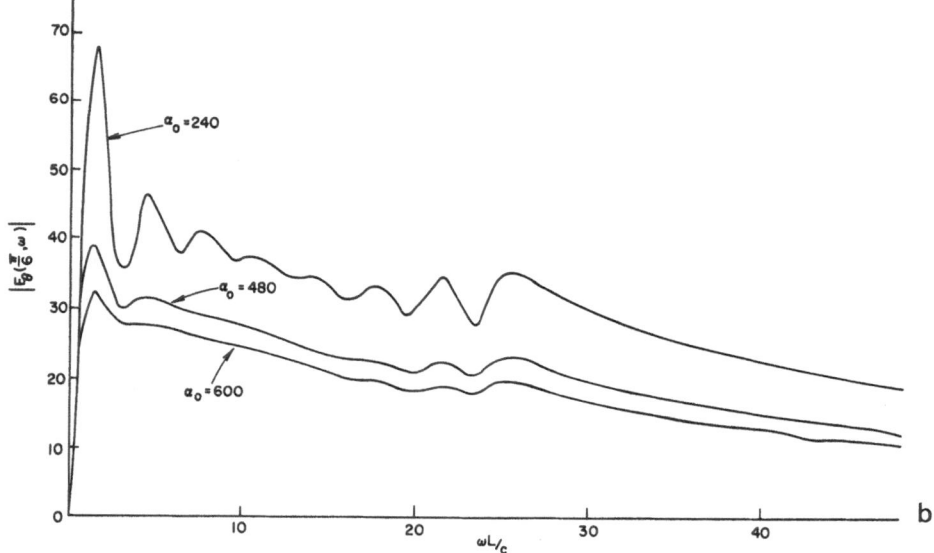

Fig. 4.10a and b. Envelope of $|E_\theta(\theta, \omega)|$ as a function of normalized frequency with α_0 as the parameter. $L/a = 156$ (after Liu and Sengupta [4.22]). a) $|E_\theta(\pi/2, \omega)|$. b) $|E_\theta(\pi/6, \omega)|$

peak appears to be independent of θ. As θ decreases from 90°, the minor peaks in $E_\theta(\theta,\omega)$ become appreciable.

b) As α_0 increases, the low frequency content of $|E_\theta(\theta,\omega)|$ and the initial peak in $|E_\theta(\theta,\omega)|$ decreases significantly. The rate of decrease of $|E_\theta(\theta,\omega)|$ in the high frequency end appears to be almost independent of α_0.

c) In the broadside direction, the position of the initial peak in $|E_\theta(\pi/2,\omega)|$ increases with an increase in the loading parameter α_0. For example, from Fig. 4.10a it is found that the initial peaks are located at $\omega L/c \simeq 1.56$, 2.4 and 2.6 for $\alpha_0 = 240$, 480 and 600, respectively.

d) The decrease of L/a tends to make the minor peaks in $|E_\theta(\theta,\omega)|$ less pronounced.

The above observations will have implications in the corresponding time dependent wave forms.

4.5.4 Time-Dependent Radiated Wave Form

Figures 4.11 and 4.12 show the time dependent wave forms in two different directions radiated by loaded linear antennas for $L/a = 156$ and for various values of the loading parameter α_0. In each case $L = 1$ meter, so that $\tau = 3.33$ ns and the width of the input rectangular voltage pulse $T = 20.94$ ns. It has been found [4.33] that for $T \gtrsim 3\tau$, the time dependent radiated wave forms due to the discontinuities at the two ends of the input pulse do not interfere with each other. The numerical

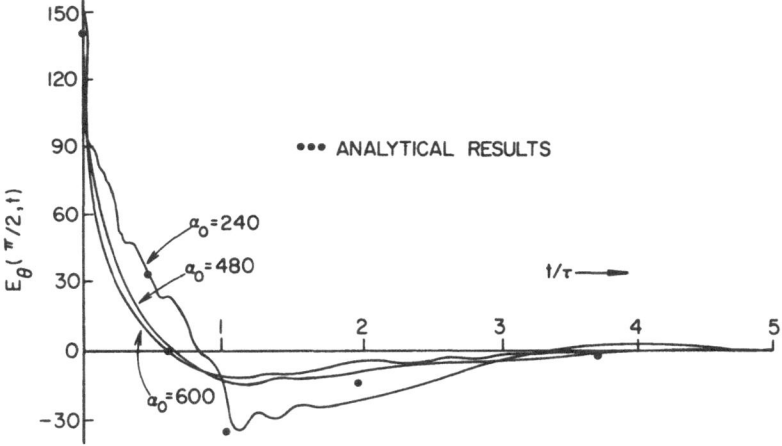

Fig. 4.11. $E_\theta(\pi/2,t)$ as a function of t/π with α_0 as the parameter for step voltage excitation. $\tau = 3.33$ ns, $T/\tau \simeq 6$ and $L/a = 156$ (after LIU and SENGUPTA [4.22])

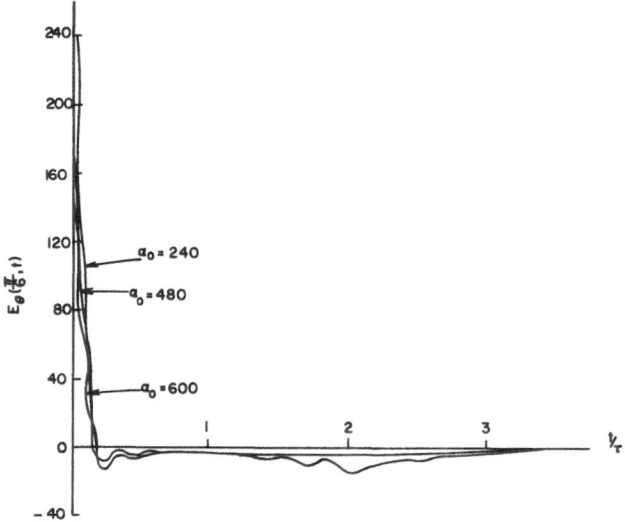

Fig. 4.12. $E_\theta(\pi/6, t)$ as a function of t/τ with α_0 as the parameter for step voltage excitation. $\tau = 3.33$ ns, $T/\tau \simeq 6$ and $L/a = 156$ (after Liu and Sengupta [4.22])

calculations were truncated at those t/τ values for which the field was sufficiently small. Hence the results shown here for $T = 20.94$ ns may be identified with the transient wave forms radiated by the antenna for step voltage excitation. It is anticipated that for the unloaded antenna T must be larger than the value used here for proper approximations.

The undulations in the wave form for $\alpha_0 = 240$ (for example, Fig. 4.11) are attributed to the existence of reflected current waves on the antenna. In general, the existence of reflection effects causes the wave form to cross the zero axis more than once, as can be seen in the $\alpha_0 = 240$ wave form in Fig. 4.11. For α_0 greater than the critical value, the radiated wave form has only one zero crossing. It has been found that for a given angle θ, the increase of α_0 has the following effects on the radiated wave forms:

a) Decreases the initial amplitude $E_\theta(\theta, t)$. In the asymptotic results obtained from analytical studies, it is found [4.17] that $E_\theta(\theta, t) \propto 1/t^{1/2}$ as $t \to 0$.

b) Increases the rate of decay of $E_\theta(\theta, t)$ after the initial rise.

c) Decreases the first zero crossing time t_0 and consequently decreases pulse width.

d) Decreases the amplitude $E_\theta(\theta, t)$ for $t > t_0/\tau$.

For a given value of the loading parameter α_0, the decrease of the observation angle θ from the broadside direction ($\theta = \pi/2$) has the following effects on the radiated wave forms:

a) Increases the rate of decrease of $E_\theta(\theta, t)$ from its initial amplitude and decreases the zero crossing time t_0. Consequently, the pulse width decreases.

b) Increases the initial amplitude of $E_\theta(\theta, t)$.

The shape of the wave form radiated by a non-reflecting antenna can be adjusted by adjusting the zero crossing time t_0. For this purpose it is of interest to study the variation of t_0 as a function of the various physical parameters of the antenna. Figure 4.13 shows the normalized zero crossing time as a function of loading with the observation angle θ as the parameter. Notice that results are shown for values of $L/a = 156$ and 100. As far as the zero crossing time is concerned, it is found from Fig. 4.13 that the decrease of θ has similar effects on t_0 as the increase of α_0, i.e., t_0 decreases with a decrease in θ. In the broadside direction, t_0 is found to decrease with the decrease of L/a. In this direction, the spread between the t_0 values, for the two values of L/a shown, increases with an increase of α_0.

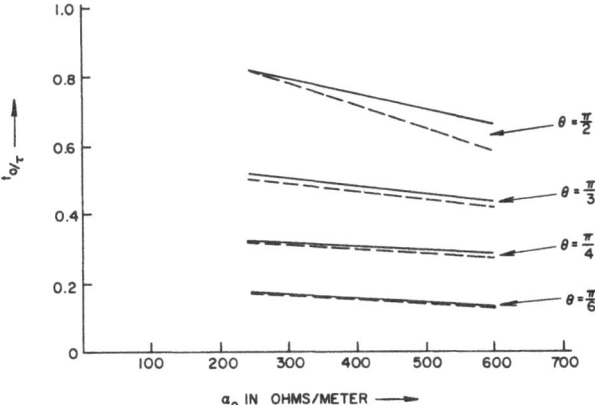

Fig. 4.13. t_0/τ as a function of α_0 with θ as the parameter. $\tau = 3.33$ ns, $T = 20.94$ ns; —— $L/a = 156$, ---- $L/a = 100$ (after LIU and SENGUPTA [4.22, 23])

Assuming the current distribution on the harmonically excited antenna as given by (4.66), it is possible to obtain a closed form expression for the radiated wave form with the help of (4.39) through (4.46). The analytical results obtained in this way for the non-reflectively loaded antenna with $\alpha_0 = 690$ are also exhibited in Fig. 4.11 for comparison. Although the analytical results diplayed are for a different value of α_0, it is possible to make some general comments by studying these results along with the numerical results. Physical considerations dictate that the initial rate of decay of the field should decrease with an increase in α_0 and also that the magnitude of the field at late time should decrease with increasing α_0. However, it appears from Fig. 4.11 that the analytical results predict slightly slower decay of $E_\theta(\pi/2, t)$ for $t < t_0$ than the numerically obtained values for $\alpha_0 = 480$ and 600. For $t > t_0$ the approximate theory predicts a larger value of $E_\theta(\pi/2, t)$. These discrepancies may be attributed to the fact that the theoretical model assumes a linearly decaying current amplitude distribution on the harmonically excited antenna, whereas the numerically obtained results discussed earlier indicate that the amplitude of the current distribution on the antenna is in general approximately exponentially decaying.

The early time behavior of $E_\theta(\theta, t)$ is compared with the asymptotic results [4.17] for uniformly loaded infinite dipole antennas. It is assumed that the infinite dipole antenna is loaded uniformly in such a manner that the internal impedance of the antenna is α_0 [Ω/m]. We compare the early time behavior of the wave form radiated by this infinite antenna with that of the wave form produced by the non-reflectively loaded finite antenna for $t \ll \tau$. It is assumed that the radius a is the same for both antennas. The results are shown in Fig. 4.14 which indicates

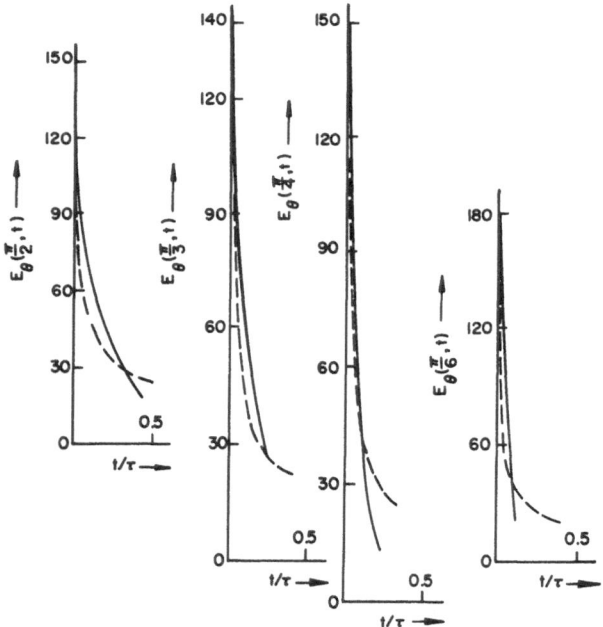

Fig. 4.14. Early time behavior of $E_\theta(\theta, t)$ as a function of time. $\tau = 3.33$ ns, $T = 20.94$ ns, $\alpha_0 = 480$, $L/a = 156$; —— numerical; ---- asymptotic results of Latham and Lee [4.17] (after Liu and Sengupta [4.22, 23])

that the theory in [4.17] predicts faster decay of $E_\theta(\theta, t)$ for early times. With the assumption that the antenna is loaded uniformly with α_0, the decay rate of the field for early times should have been slower. This discrepancy indicates that the non-uniformly loaded finite antenna may not be approximated by a uniformly loaded infinite antenna for $t \ll \tau$, an assumption usually made for the uniform case.

4.6 Space-Time Integral Equation Method of Solution

The space-time integral equation is derived for the time-dependent current distribution on linear antennas excited by time-dependent slice generators. For general three-dimensional structures the derivation of the same equation may be found in [4.27] (see also Chapt. 2). From the knowledge of the time-dependent current, formal expressions are given for the various time-dependent field quantities. Numerical procedures for obtaining quantitative results are also discussed. This methods of obtaining solutions to transient electromagnetic problems was first reported in [4.39].

4.6.1 Fundamental Relations

For a given current and charge density distribution $J(r, t)$ and $\varrho(r, t)$ within a specified volume V', the fields produced at any field point in space may be obtained with the help of retarded vector and scalar potentials [4.40]. These are given by

$$A(r, t) = \frac{\mu_0}{4\pi} \int_{V'} J(r', t') R^{-1} dv' \tag{4.80}$$

$$\Phi(r, t) = \frac{1}{4\pi\varepsilon_0} \int_{V'} \varrho(r', t') R^{-1} dv', \tag{4.81}$$

where r is the position vector of the field point whose spherical coordinates are (r, θ, ϕ), r' is the position vector of the source point, $R = |r - r'|$ is the distance between the field and the source points, and $t' = t - R/c$ is the retarded time. The above two potentials are related to each other by the Lorentz condition

$$\nabla \cdot A(r, t) + (1/c^2) \partial \Phi(r, t)/\partial t = 0. \tag{4.82}$$

The source current and charge densities together satisfy the equation of continuity

$$\nabla \cdot J(r,t) + \partial \varrho(r,t)/\partial t = 0. \tag{4.83}$$

The electric and magnetic fields produced by the above source distribution are given by the following

$$E(r,t) = -\partial A(r,t)/\partial t - \nabla \Phi(r,t), \tag{4.84}$$

$$H(r,t) = (1/\mu_0) \nabla \times A(r,t). \tag{4.85}$$

4.6.2 Space-Time Integral Equation for the Current Distribution

Consider a linear antenna of length $2L$ excited at the center by a delta gap as shown in Fig. 4.1. For the present we do not characterize the source completely except for the fact that it is a time dependent slice generator [4.41]. Our purpose here is to develop an integral equation for the current distribution on the antenna (see also Subsect. 2.4.3). Throughout this section it is assumed that the field point is located on the antenna surface. We assume that the antenna is sufficiently thin so that the retarded potentials at a point on the surface of the antenna may be obtained from (4.80) and (4.81) in the following forms

$$A(r_s,t) = z_0 \frac{\mu_0}{4\pi} \int_{-L}^{L} \frac{I(z',t_s')}{R_s} \, dz', \tag{4.86}$$

$$\Phi(r_s,t) = \frac{1}{4\pi\varepsilon_0} \int_{-L}^{L} \frac{q(z',t_s')}{R_s} \, dz', \tag{4.87}$$

where $I(z',t')$, $q(z',t')$ are the total current and charge densities per unit length on the antenna at $z = z'$, and

$$R_s = [(z-z')^2 + a^2]^{1/2}. \tag{4.88}$$

The equation of continuity (4.68) takes the following form for the present case

$$\partial I(z,t)/\partial z + \partial q(z,t)/\partial t = 0 \tag{4.89}$$

The total electric field at any point on the surface of the antenna is

$$E_T(r_s,t) = E_e(r_s,t) + E_a(r_s,t), \tag{4.90}$$

where $E_e(r_s, t)$ is the field due to the externally impressed source only, and $E_a(r_s, t)$ is the field due to the antenna current $I(z', t)$ only. In terms of the retarded potentials (4.90) may be written as

$$E_T(r, t) = E_e(r, t) - \partial A(r, t)/\partial t - \nabla \Phi(r, t). \tag{4.91}$$

After differentiating (4.91) with respect to time and making use of the Lorentz condition, we obtain the following relation valid at a point on the antenna surface

$$\partial E_T(r_s, t)/\partial t = \partial E_e(r_s, t)/\partial t - \partial^2 A(r_s, t)/\partial t^2 + c^2 \nabla [\nabla \cdot A(r, t)], \quad (4.92)$$

Taking the z-component of (4.92) and making use of (4.86) we obtain the following space-time integral equation of the current distribution on the antenna

$$[\partial^2/\partial z^2 - (1/c^2)\partial^2/\partial t^2]\frac{\mu_0}{4\pi} \int_{-L}^{L} I(z', t') R_s^{-1} dz'$$
$$= -(1/c^2)(\partial E_e/\partial t)_z + (1/c^2)(\partial E_T/\partial t)_z \quad (4.93)$$

For a perfectly conducting antenna, $E_T(r_s, t)_z = 0$ and if it is assumed that $E_e(z, t) = z_0 V_0 \delta(z) V(t)$, where $V(t)$ is the time variation of the input voltage, then we obtain the following space-time integral equation for the current distribution $I(z, t)$:

$$[\partial^2/\partial z^2 - (1/c^2)\partial^2/\partial t^2](\mu_0/4\pi) \int_{-L}^{L} I(z', t') R_s^{-1} dz'$$
$$= -[V_0 \delta(z)/c^2][\partial V(t)/\partial t]. \quad (4.94)$$

Note that if $V(t)$ is a unit step function in time, then $\partial V(t)/\partial t = \delta(t)$ on the right hand side of (4.79).

If the antenna is continuously loaded with $Z^i(z)$ ohms per unit length, then

$$E_T(z, t)|_z = z_0 Z^i(z) I(z, t)$$

Under this condition we obtain the following integral equation

$$[\partial^2/\partial z^2 - (1/c^2)\partial^2/\partial t^2](\mu_0/4\pi) \int_{-L}^{L} I(z', t')/R_s dz'$$
$$= [Z^i(z)/c^2]\partial I(z, t)/\partial t - V_0[\delta(z)/c^2]\delta v(t)/\partial t. \quad (4.95)$$

4.6.3 Field Expressions

In this Subsection it is assumed that the current distribution $I(z,t)$ on the linear antenna is known and is constrained on the z-axis. Expressions are derived for the fields produced. The coordinate system used is as shown in Fig. 4.1. The vector potential produced by the current $I(z,t)$ at the field point $P(r,\theta,\phi)$ is

$$A(r,t) = z_0(\mu_0/4\pi) \int_{-L}^{L} I(z',t')R^{-1}dz', \qquad (4.96)$$

where $t' = t - (R/c)$ and the other notations are as explained in Fig. 4.1.

With the help of (4.85) and (4.96) we obtain the following expression for the magnetic field $H(r,t)$ at P

$$H(P) = (z_0 \times R_0)(1/4\pi) \int_{-L}^{L} [1/R^2)I(z',t') + (1/Rc)(\partial/\partial t')I(z',t')] dz' \qquad (4.97)$$

where R_0 is the unit vector in the direction of R. If the field point P is at a large distance from the antenna then we can approximate

$$R \simeq r - z' \cos\theta$$

$$R_0 \simeq r_0.$$

Under this condition we obtain the following from (4.97):

$$H(P) = \phi_0 \sin\theta(4\pi)^{-1} \int_{-L}^{L} [(1/r^2)I(z',t') + (1/rc)(\partial/\partial t')I(z',t')] dz' \quad (4.98)$$

where

$$t' = t - (r/c) + (z'c^{-1}\cos\theta). \qquad (4.99)$$

In the far zone the term depending on $1/r^2$ is neglected and we obtain the following expression for the far-zone magnetic field

$$H(P) = \phi_0(4\pi rc)^{-1}\sin\theta \int_{-L}^{L} (\partial/\partial t')I(z',t')dz', \qquad (4.100)$$

with t' given by (4.99).

It can be shown the time dependent electric field $E(r,t)$ is given by

$$E(r,t) = -z_0(\mu_0/4\pi) \int_{-L}^{L} (\partial/\partial t')[I(z',t')/R]dz'$$

$$+ (1/4\pi\varepsilon_0)R_0 \int_{-L}^{L} \left[\frac{q(z',t')}{R^2} + (1/Rc)\frac{\partial q(z',t')}{\partial t'} \right] dz'. \qquad (4.101)$$

Using the far-field approximations we obtain the following for the electric field in the far zone of the antenna,

$$E(r,t) \simeq \theta_0 \mu_0 (4\pi r)^{-1} \sin\theta \int_{-L}^{L} (\partial/\partial t')I(z',t')dz', \qquad (4.102)$$

with t' given by (4.99).

4.6.4 Numerical Solution of the Space-Time Integral Equation

To obtain the time dependent field produced by the linear antenna one must know the time-dependent current distribution on the antenna. This is obtained by solving numerically the space-time integral equation discussed in Subsect. 4.6.2. Here we give a brief outline of the numerical procedure which may be used to solve the integral equation. The method described here is essentially the same as in [4.23] (see also Chapt. 2). For this purpose it is found convenient to express the time coordinate in terms of light meters, i.e., use the transformation $t_1 = ct$, and we write the integral equation (4.95) as

$$\Box^2 \left\{ (4\pi)^{-1} \int_{-L}^{L} [I(z',t_1')/R_s]dz' \right\}$$

$$= \eta_0^{-1} Z^i(z)[\partial I(z,t_1)/\partial t_1] - \eta_0^{-1}(\partial E^e/\partial t_1)_z, \qquad (4.103)$$

where

$$\Box^2 = \partial^2/\partial z^2 - (1/c^2)\partial^2/\partial t^2 = \partial^2/\partial z^2 - \partial^2/\partial t_1^2, \quad t_1 = ct, \qquad (4.104)$$

$$t_1' = t_1 - R_s \qquad (4.105)$$

In terms of the modified units, the far zone electric and magnetic fields produced by the antenna are given by

$$E(r,t_1)=\theta_0\eta_0(4\pi r)^{-1}\sin\theta\int_{-L}^{L}(\partial/\partial t_1')I(z',t_1')dz' \qquad (4.106)$$

$$H(r,t_1)=\phi_0(4\pi r)^{-1}\sin\theta\int_{-L}^{L}(\partial/\partial t_1')I(z',t_1')dz' , \qquad (4.107)$$

where

$$t_1'=t_1-r+z'\cos\theta . \qquad (4.108)$$

Equation (4.103) is to be solved for the current distribution $I(z,t_1)$ subject to the end condition

$$I(\pm L,t_1)=0 . \qquad (4.109)$$

For the purpose of numerical evaluation we rearrange the integral on the left-hand side of (4.103) as follows

$$1/4\pi\int_{-L}^{L}[I(z',t_1')/R_s]dz'=(1/4\pi\left(\int_{-L}^{z-\Delta z}+\int_{z-\Delta z}^{z+\Delta z}+\int_{z+\Delta z}^{L}\right)[I(z',t_1')/R_s]dz' \qquad (4.110)$$

where R_s is given by (4.88). For sufficiently small Δz we can write

$$(4\pi)^{-1}\int_{z-\Delta z}^{z+\Delta z}[I(z',t_1')/R_s]dz'\simeq\xi I(z,t_1), \qquad (4.111)$$

where

$$\xi=(2\pi)^{-1}\ln\{[\Delta z+(\Delta z^2+a^2)^{1/2}]a^{-1}\}. \qquad (4.112)$$

After using (4.110) through (4.112), the space-time integral equation (4.103) can be written in the following form

$$\Box^2 I(z,t_1)-\eta_0^{-1}Z^i(z)\partial I(z,t_1)/\partial t_1$$
$$=-\xi^{-1}\left\{1/\eta_0(\partial E^e/\partial t_1)_z+(4\pi)^{-1}\Box^2\left(\int_{-L}^{z-\Delta z}+\int_{z+\Delta z}^{L}\right)[I(z',t_1')/R_s]dz'\right\} \qquad (4.113)$$

where

$$t_1'=t_1-R_s \qquad (4.114)$$

The numerical scheme developed in [4.23] to obtain $I(z, t_1)$ from (4.113) is accomplished by treating the right-hand side as known. This is possible in the present case, because the right hand side of (4.113) contains the exciting field which is known and the antenna currents delayed in time (see also Subsect. 2.3.1).

Equation (4.113) can be solved numerically on a digital computer by marching on time (see also Subsect. 2.4.3). For this purpose the wave operator $\Box^2$ and the time differentiation are approximated by the following difference equations [4.43, 44],

$$\Box^2 I(z_j, t_{1j}) \tag{4.115}$$
$$= (I_{i-1,j-1} - 2I_{i,j-1} + I_{i+1,j-1})/(\Delta z)^2 - (I_{i,j-2} - 2I_{i,j-1} + I_{i,j})/\Delta t_1)^2$$

$$\partial I(z_i, t_{1j})/\partial t_1 = (I_{i,j} - I_{i,j-1})/\Delta t_1 \tag{4.116}$$

where

$$I_{i,j} = I(z_i, t_{ij})$$

$$\Delta z = \text{space increment}$$

$$\Delta t_1 = \text{time increment}.$$

With the above approximation, (4.113) reduces to

$$[(I_{i-1,j-1} - 2I_{i,j-1} + I_{i+1,j-1})(\Delta z)^{-2} - (I_{i,i-2} - 2I_{i,j-1} + I_{i,j})(\Delta t_1)^{-2}]$$
$$- \eta_0^{-1} \xi^{-1} Z^i(z_i)(I_{i,j} - I_{i,j-1})(\Delta t_1)^{-1} = -F_{i,j-1} \tag{4.117}$$

where

$$F_{i,j} = \xi^{-1} [\eta_0^{-1}(\partial E^e/\partial t)|_{z_i,t_{1j}} + (4\pi)^{-1} \Box^2 g_{ij}], \tag{4.118}$$

$$g_{i,j} = \int_{-L}^{z_{i-1}} [I(z', t_1')/R_s] dz' + \int_{z_{i+1}}^{L} [I(z', t_1')/R_s] dz'. \tag{4.119}$$

For simplicity of computational procedure, (4.117) may be rearranged as follows

$$[(\Delta t_1)^2/(\Delta z)^2][I_{i-1,j-1} - 2I_{i,j-1} + I_{i+1,j-1}] - [I_{i,j-2} - 2I_{j,i-1} + I_{i,j}]$$
$$- \eta_0^{-1} \alpha^{-1} Z^i(z_i) I_{i,j} \Delta t_1 + \eta_0^{-1} \alpha^{-1} Z^i(z_i) I_{i,j-1} \Delta t_1 = -(\Delta t_1)^2 F_{i,j-1}. \tag{4.120}$$

For the convergence of the computation procedure it is necessary that $\Delta t_1 \leq \Delta z$ (i.e., $\Delta t < \Delta z/c$) [4.43, 44].

The recurrence relation for the current on the wire is obtained from (4.120) as

$$I_{i,j}[1 + \eta_0^{-1}\xi^{-1}Z^i(z_i)\Delta t_1] = (\Delta t_1)^2 F_{i,k-1} + (\Delta z)^{-2}(\Delta t_1)^2$$

$$\times [I_{i-1,j-1} - 2I_{i,j-1} + I_{i+1,j-1}] - (I_{i,j-2} - 2I_{i,j-1}). \quad (4.121)$$

Equation (4.121) is solved on a digital computer for the current density by simply marching on time. The computer program is available in [4.23].

4.6.5 Some Typical Results

Numerical results obtained by this method for the transient responses of linear antennas are discussed in [4.23] and [4.27]. Figure 4.15a shows the driving point current as a function of the normalized time t/τ ($\tau = L/c$) for a center-fed dipole antenna of length $2L$ and fed with a Gaussian-time-dependent source (see also Figs. 2.7–9). The peaks in the input current appear to occur at approximately odd integer multiples of τ, the transient time from the center of the antenna to one end. These peaks are due to reflections from the antenna end points. The input admittance of the antenna may be obtained from the ratio of the Fourier transform of the driving point current to that of the source voltage. Figure 4.15b shows the input conductance and suspectance of the same antenna obtained from the current shown in Fig. 4.15a. For comparison the frequency domain results obtained from [4.45] are also shown in Fig. 4.15b. The broadside radiated wave form produced by the antenna is shown in Fig. 4.16. Here again, the effects of reflections from the ends are quite clear. For further discussion of results see [4.27] and Section 2.5.

4.7 Time-Domain Integral Equation

The time-dependent current distribution on a thin linear antenna can also be found by an alternative approach which solves an integral equation of Hallen type in the time domain [4.25]. The space-time integral equation discussed in Section 4.6 contains time and/or space derivatives. The numerical representations of these derivatives are not always satisfactory and these methods, as presented, involve complicated

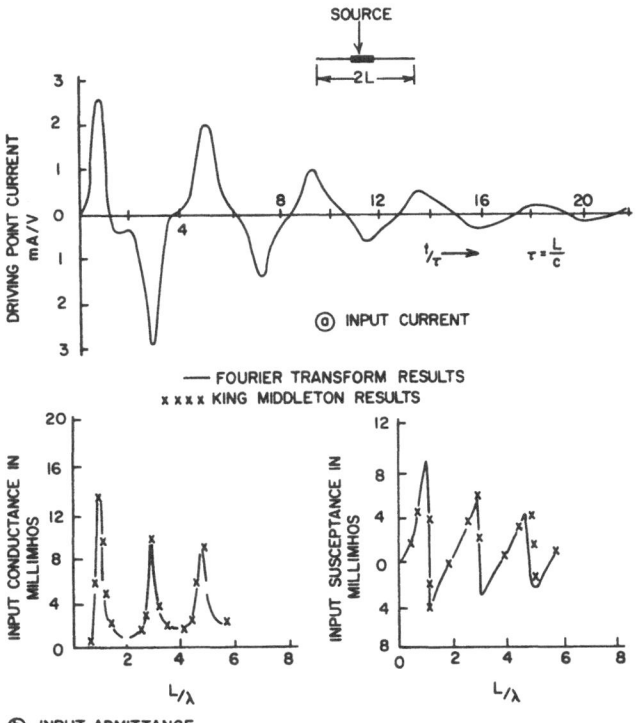

Fig. 4.15a and b. Driving point current and input admittance of a linear dipole antenna excited by a Gaussian source. $L/a \simeq 75$, $\tau = 1.67$ ns, $V_g(t) = \exp[a^2(t - t_{max})^2]$ with $a = 1.5 \times 10^9$, $t_{max} = 1.43 \times 10^{-9}$ s (after MILLER [4.27])

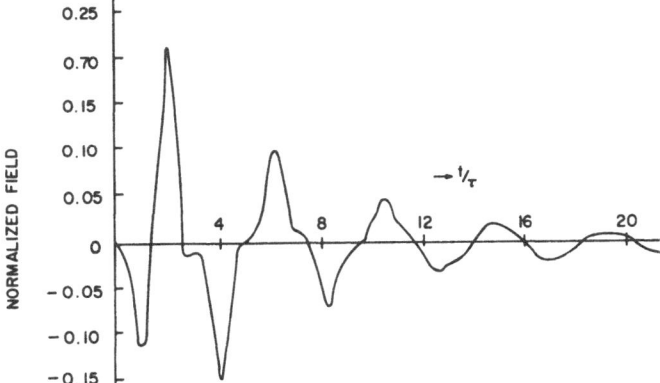

Fig. 4.16. Time dependent broadside far field radiated by the linear dipole antenna excited by the Gaussian source of Fig. 4.15 (after MILLER [4.27])

numerical procedures. The Hallen-type time-domain integral equation does not contain any space-time derivatives of the unknown in the integral operator. As a result it does possess some definite advantages over other methods.

For an unloaded thin wire linear antenna of length $2L$ and radius a, the time-dependent current distribution satisfies the following integral equation [4.25]

$$\int_0^{2L} (4\pi)^{-1} I(z',t-|z-z'|/c)[(z-z')^2+a^2]^{-1/2}\,dz'$$

$$=(1/2\eta_0)\int_0^L E_{\text{inc}}(z',t-|z-z'|/c)\,dz'+f_1(ct-z)+f_2(ct+z), \qquad (4.122)$$

where $E_{\text{inc}}(z,t)$ is the z-directed incident field, and $f_1(ct-z)$, $f_2(ct+z)$ are the homogeneous solutions of the equation

$$\partial^2 A_z(z,t)/\partial z^2 -(1/c^2)\partial^2 A_z(z,t)/\partial t^2 = -(1/c^2)\partial E_{\text{inc}}(z,t)/\partial t, \qquad (4.123)$$

where $A_z(z,t)$ is the magnetic vector potential at the surface of the perfectly conducting antenna. The solutions f_1, f_2 are valid on the surface of the antenna subject to the boundary condition $I(0,t)=I(2L,t)=0$.

Equation (4.122) indicates that for a specific pair of values (z,t_0) the current $I(z',t_0-|z-z'|/c)$ lies on the straight lines

$$\left.\begin{array}{c} t-(z-z')/c=t_0 \\ t+(z-z')/c=t_0 \end{array}\right\}, \qquad (4.124)$$

in the (z',t) plane, as shown in Fig. 4.17. For a discrete set of parameters t_0, these straight lines form two families of trajectories, α and β, known as the characteristic curves [4.46]. On the α-family of curves the current waves propagate with velocity c along the wire in the $+z$ direction while on the β-family of curves they propagate with the same velocity in the negative z direction. The solutions $f_1(ct-z)$ and $f_2(ct+z)$ are invariant along α and β curves, respectively. These two functions in the equations account for the multiple reflection of the current at the two ends of the wire.

The current at any time t may be found from (4.122) if the current and f_1, f_2 along the α and β characteristic curves are known prior to time t. For this purpose, the characteristic curves in Fig. 4.17 are extended beyond the line at $t=0$. Using the fact $f_1(ct-z)=f_2(ct+z)=0$ for $t\le 0$, and that f_1, f_2 are invariant along these characteristic curves,

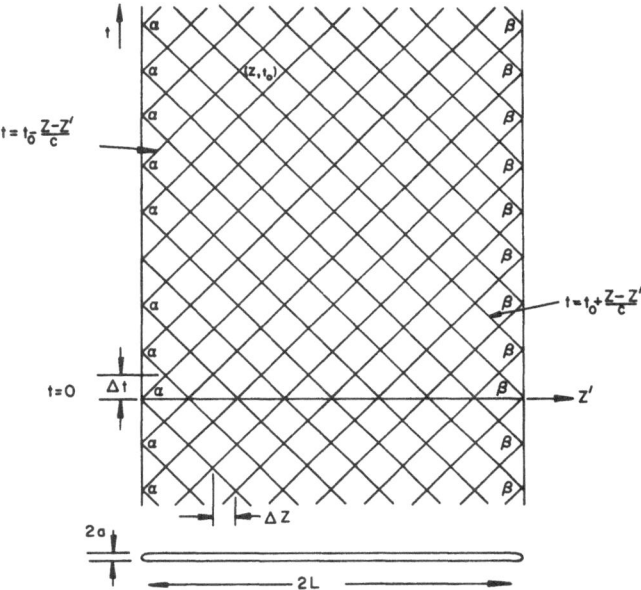

Fig. 4.17. Diagram of $z'-t$ showing the two families of α and β characteristics (after LIU and MEI [4.25])

we find that these two functions vanish as long as the (z, t) pair is on a characteristic curve which extends into the region $t \leq 0$. To find the current at $t = \Delta t$, it is assumed at first that the antenna is unexcited, i.e., $I(z, 0) = 0$. Thus the current at $t = \Delta t$ may be found from (4.122). At $t = 2\Delta t$, since $I(0, t) = I(L, t) = 0$ we find $f_1(2c\Delta t)$ and $f_2(2c\Delta t + 2L)$ on the outgoing characteristic curves at $z = 0$ and $z = 2L$, respectively. Using similar steps, the current may be found for other times. The computation is a step-by-step time marching process, based on the condition that the currents prior to the point under evaluation are known. Thus if we can perform the calculation for one time line, we are able to compute for the current at all subsequent times. Numerical approximation of the integrals in (4.122) can be done by standard moment methods [4.34].

Figure 4.18 shows the driving point current response of a center-fed antenna excited by a unit time step voltage for the case with $L/a = 74.2$. The reflection effects from the two ends, the rapid decay of the high frequency components and the eventual no-current flow at late time are clearly evident. Figure 4.19 shows the normalized far field responses of the above antenna for $\theta = 15$ to $90°$ at $15°$ intervals. For further results obtained by this method, see [4.25].

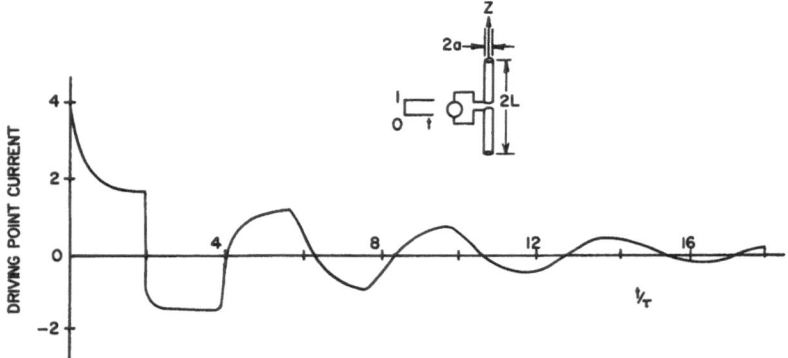

Fig. 4.18. Driving point current for a center fed linear antenna under a unit time step voltage excitation. $L/a = 74.2$, $\tau = L/c = 1.67$ ns (after Liu and Mei [4.25])

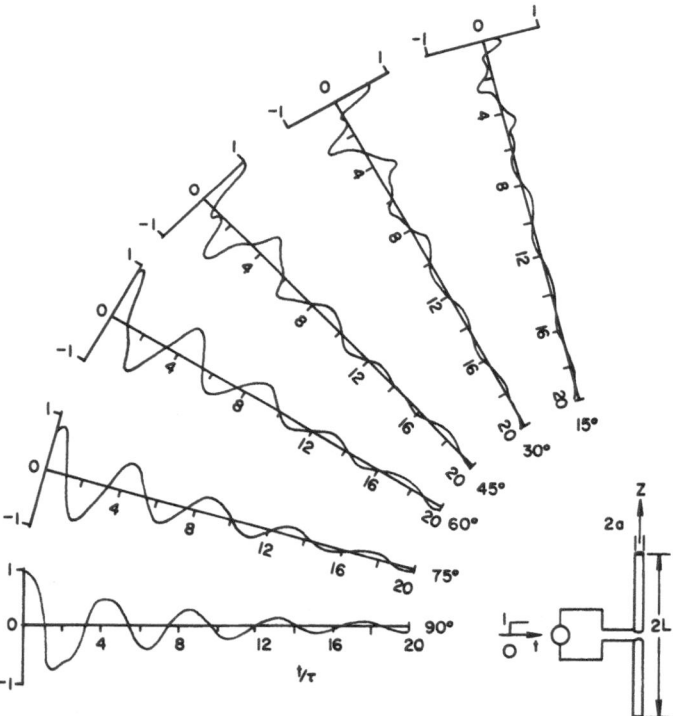

Fig. 4.19. Normalized far field wave forms radiated by a center fed linear antenna under a unit time step voltage excitation. $L/a = 74.2$, $\tau = 1.67$ ns (after Liu and Mei [4.25])

4.8 Direct Solution in the Time-Domain

The linear antenna problem can also be solved numerically directly in the time domain without using the space-time or the time domain integral equation discussed earlier. This is done by applying the moment methods directly to solve the fundamental equations (4.84) through (4.89) subject to the appropriate boundary conditions [4.26]. Equations (4.84) and (4.89) are approximated by finite difference operators. Equations (4.86) and (4.87) are satisfied through the use of series representations for the currents and charge as functions of time and space. The expansions satisfy the initial and boundary conditions of the problem. The solution starts at $t=0$ and iterates forward in time. The impressed electric field exists only at the driving point and is zero elsewhere along the antenna.

For step voltage excited antenna with $L/a=72.4$, the driving point current as a function of normalized time is shown in Fig. 4.20. The

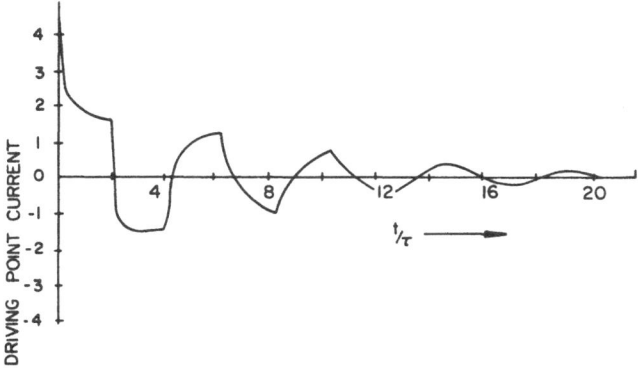

Fig. 4.20. Driving point current of a center fed linear dipole antenna of length $2L$ and excited by a unit time step voltage. $L/a=74.2$, $\tau=L/c=1.67$ ns (after SAYRE and HARRINGTON [4.26])

solution corresponds to using 20 subsections of the antenna wire and 20 increments per transit time $(2L/c)$. The initial value of the current is in good agreement with that calculated for an infinitely large linear antenna [4.13]. The reflection effects from the two ends, the rapid decay of high frequency components and the eventual no-current flow at late times are clearly seen in Fig. 4.20. The general periodic behavior of the currents is evident, with the period corresponding closely to the fundamental mode of the center dipole. The input admittance of the antenna calculated from the current is shown in Fig. 4.21. The

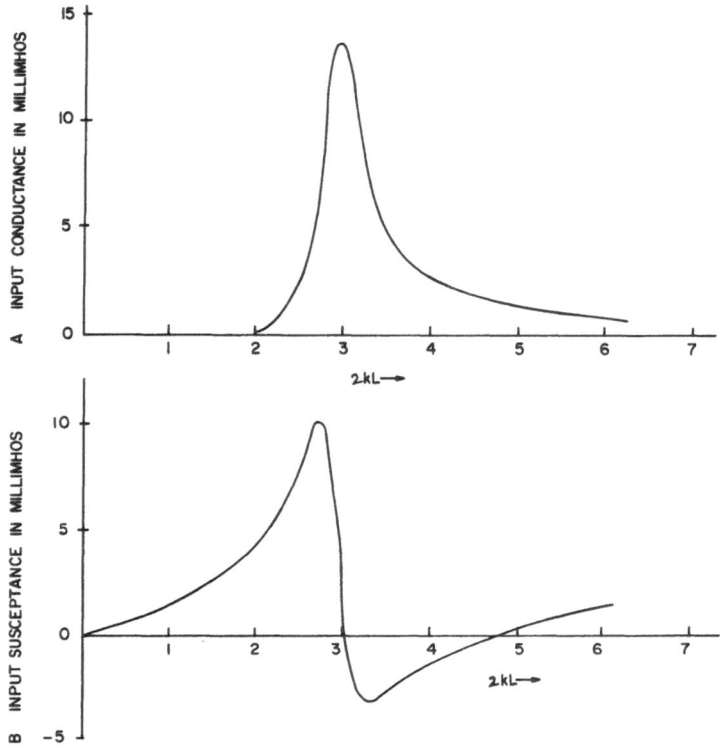

Fig. 4.21 A and B. Driving point admittance calculated from the input current shown in Fig 4.20 (after Sayre and Harrington [4.26])

values agree closely with results found in [Ref. 4.34, p. 72]. The peak at $kL \simeq 1.5$ is the fundamental antenna mode and corresponds closely in the time domain to twice the transit time from the feed point to the end of the antenna. Further discussion of the results may be found in [4.26].

4.9 Infinitely Long Linear Antenna

The infinitely long linear antenna excited at the center with a slice generator is one of the very few antennas whose transient radiation can be obtained exactly by analytical means [4.13, 14, 17]. Although such an antenna is impractical, the results of its transient analysis may be applied to finite length dipoles under some special conditions.

For example, if the maximum linear dimension of the excitation region of the antenna is small compared with its length and if the excitation voltage is zero until some initial instant, there will be a significant time interval at any point in space during which the field will be very nearly equal to the field radiated by an infinite cylindrical antenna with a slice generator excited by the same voltage wave form. In particular, at a distant point along a line through the gap and broadside to the axis of the antenna, it will be impossible to detect the finiteness of the length of the antenna until at least at time interval L/c after the arrival of the front edge of the pulse, where L is the minimum length from the gap to one end of the dipole. The discussion of the transient radiation from an infinitely long linear antenna, given below, is based on that given in [4.17].

4.9.1 H-field

Consider a perfectly conducting infinite cylindrical antenna of radius a excited by a slice generator, as shown in Fig. 4.1, with $L=\infty$. If the slice generator produces a time independent field $E_z(s)=-V(s)\delta(z)$ on its surface, then it can be shown that the time independent magnetic field produced by the antenna at a point $P(\varrho,z)$ is given by [4.42]

$$H_\phi(\varrho,z,p)=-(2\pi\eta_0)^{-1}pV(pc)\int_{-\infty}^{\infty}K_1(\bar{v}\varrho)/[\bar{v}K_0(\bar{v}a)]\exp(i\zeta z)d\zeta$$

$$(4.125)$$

where $pc=-i\omega=s$ is the Laplace transform variable, c being the velocity of light in free space, $\bar{v}=(p^2+\zeta^2)^{1/2}$, $V(pc)=V(s)$ is the Laplace transform of the input voltage, and K_0, K_1 are the usual notations for the modified Bessel functions of the second kind. The path of integration and the branch of $\bar{v}$ involved in (4.125) are shown in Fig. 4.22.

When the excitation is a step function of voltage V_0 such that $V(t)=V_0U(t)$ we have $V(s)=V(pc)=V_0/pc$. For this excitation the time dependent magnetic field is obtained by taking the inverse Laplace transform of (4.125) and is given by

$$H_\phi(\varrho,z,t)=\eta_0^{-1}V_0(2\pi i)^{-1}\int_{C_P}\exp(pct)dp(2\pi)^{-1}\int_{-\infty}^{\infty}K_1(\bar{v}\varrho)/$$
$$[\bar{v}K_0(\bar{v}a)]\exp(i\zeta z)d\zeta \quad (4.126)$$

where C_p is an integration path in the right half of the complex p-plane and runs parallel to the imaginary axis. By using the techniques of contour integration involving branch cuts of multiple-valued functions

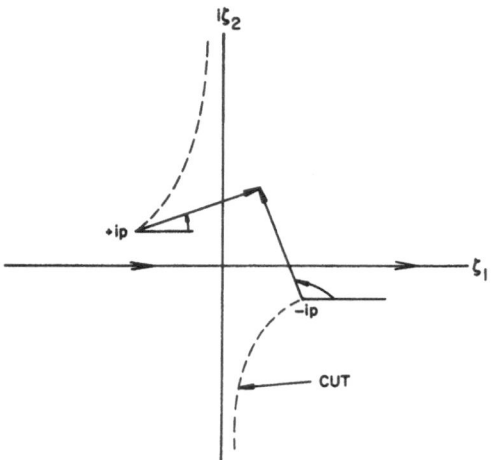

Fig. 4.22. Branch of $\bar{v}=(\zeta^2+p^2)^{1/2}$
for $\bar{v}=p$ at $\zeta=0$

it can be shown [4.17] that (4.126) may be obtained in the following form

$$(\eta_0/V_0)H_\phi(\varrho,z,t)=(\pi)^{-1}\int_0^\infty [J_1(\bar{v}\varrho)\,Y_0(\bar{v}a)-J_0(\bar{v}a)\,Y_0(\bar{v}\varrho)]$$

$$\times[J_0^2(\bar{v}a)+Y_0^2(\bar{v}a)]^{-1}J_0\{\bar{v}[(ct)^2-z^2]^{1/2}\}\,d\bar{v}$$

$$\text{for}\quad ct>z$$

$$=0\quad\text{for}\quad ct<z. \tag{4.127}$$

where J_0, Y_0 are the zeroth order Bessel functions of the first and second kind, respectively.

By definition the total current density on the cylinder is $I(z,t)=2\pi a H_\phi(a,z,t)$. Using this relation we obtain the following from (4.127)

$$I(z,t)=(4V_0)(\eta_0^{-1})\pi^{-1}\int_0^\infty J_0\{\bar{v}[(ct)^2-z^2]^{1/2}\}$$

$$\times[J_0^2(\bar{v}a)+Y_0^2(\bar{v}a)]^{-1}\bar{v}^{-1}\,d\bar{v}\quad\text{for}\quad ct>z. \tag{4.128}$$

The above result has also been obtained in a different manner in [4.13] and [4.16].

The expression (4.127) is exact, but in general it can be evaluated by numerical means only. For the purpose of numerical evaluation it has been found convenient [4.17] to express (4.126) in the following alternative dimensionless from

$$r(\eta_0/v_0)H_\phi(\varrho,z,t)=(r/a)\int_0^\infty \left[I_0(u)I_0(u)K_1(u\varrho/a)+K_0(u)I_1(u\varrho/a)\right]$$

$$\times\{K_0(u)[K_0^2(u)+\pi^2 I_0^2(u)]\}^{-1}K_0(u\tau/a)du$$

$$\text{for}\quad ct>[(\varrho-a)^2+z^2]^{1/2}\,,\tag{4.129}$$

where

$$(\tau/a)=[(ct)^2-(z^2)]^{1/2}a^{-1}$$

$$=(T^2+2\,T\,\{[(\varrho/a)-1]^2+(z/a)^2\}^{1/2}+[(\varrho/a)-1]^2)^{1/2}\,.\tag{4.130}$$

and I_0, I, are the usual notations for the modified Bessel functions of the first kind.

The integral (4.129) was numerically evaluated [4.17] as a function of T for several values of r and θ where T is defined as

$$T=\{ct-[(\varrho-a)^2+z^2]^{1/2}\}/a\,.\tag{4.131}$$

Some typical results are shown in Fig. 4.23.

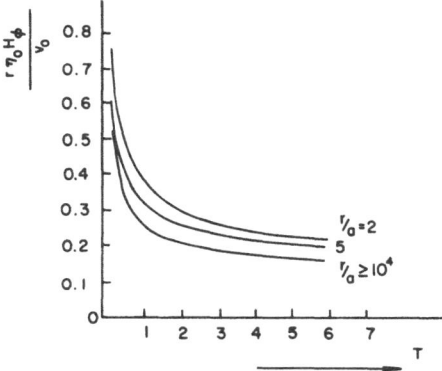

Fig. 4.23. Normalized far zone magnetic field in the broadside direction for step function excitation (after Latham and Lee [4.17])

Some limiting form of (4.129) may be derived from the asymptotic behavior of the integral for large and small T. From (4.130) we find that

$$\begin{aligned}(\tau/a)&\to\infty &&\text{as}\quad T\to\infty\,,\\(\tau/a)&\to(\varrho/a)-1 &&\text{as}\quad T\to0\,.\end{aligned}\tag{4.132}$$

Using (4.132) and the asymptotic behavior of (4.129) it can be shown [4.17] that for large T

$$r(\eta_0/V_0)H_\phi(\varrho,z,t)\sim(2\sin\theta)^{-1}[\ln(2\tau/\Gamma a)]^{-1}\,,\qquad \Gamma=1.7810\ldots\tag{4.133}$$

as $(\tau/a)\to\infty$.

For $(\tau/a) = [(\varrho/a) - 1] + \varepsilon$, with $\varepsilon \to 0$, then if $\varrho > a$ and $\varepsilon \to 0$ we obtain

$$r(\eta_0/V_0)H_\phi \sim \pi^{-1}2^{-1/2}r[\varrho(\varrho - a)]^{-1/2}\varepsilon^{-1/2}, \tag{4.134a}$$

and if $\varrho = a$ and $\varepsilon \to 0$ then

$$r(\eta_0/V_0)H_\phi \sim 2^{1/2}\pi^{-1/2}(a^2 + z^2)^{1/2}[(ct)^2 - z^2]^{-1/2}. \tag{4.134b}$$

Equations (4.134) give the small time behavior of the field.

The corresponding expressions for the field in the far zone may be obtained from (4.129). If the antenna is loaded by some constant resistance along its length such that $E_z(a, z) = -V\delta(z) + ZI(z)$ on the antenna's surface (Z being ohm meter^{-1}) then it can be shown that for step excitation

$$r(\eta_0/V_0)H_\phi(\varrho, z, t) \sim (2\sin\theta)^{-1}\int_0^\infty [I_0(x) + \beta_\theta I_1(x)]$$
$$\times \{[K_0(x) - \beta_\theta K_1(x)]^2 + \pi^2[I_0(x) + \beta_\theta I_1(x)]^2\}^{-1} \tag{4.135}$$
$$\times \exp[-x(ct - r)/(a\sin\theta)]x^{-1}dx, \quad \text{for} \quad ct > r - a\sin\theta$$
$$= 0 \qquad\qquad\qquad\qquad \text{for} \quad ct < r - a\sin\theta,$$

where

$$\beta_\theta = (2\pi a Z/\eta_0)\operatorname{cosec}\theta. \tag{4.136}$$

From (4.135) one can deduce the following asymptotic behavior

$$r(\eta_0/V_0)H_\phi \sim \pi^{-1}2^{-1/2}(\sin\theta)^{-1}(1 + \beta_\theta)^{-1}T_\theta^{-1/2} \quad \text{for} \quad T_\theta \ll 1$$
$$\sim (2\sin\theta)^{-1}(\beta_\theta^{-2}T_\theta^{-2} + \beta_\theta^{-1}T_\theta^{-3}) \qquad \text{for} \quad T_\theta \gg 1, \tag{4.137}$$

where

$$T_\theta = (ct - r)/(a\sin\theta). \tag{4.138}$$

Thus it is found that resistive loading decreases the radiation field especially in late times.

4.10 General Comments

In the preceding sections we have discussed the transient radiation and reception by linear antennas excited by slice generators having unit step or Gaussian time dependence. In general, it is found that the radiated (or received) wave form is not a faithful replica of the input wave form. The radiated wave form is distorted basically due to two reasons: the radiation mechanism from the antenna and the reflections from the two ends of the antenna. The latter effects may be considerably reduced by resistively loading the antenna. The reflectionless linear antenna is one such antenna.

Exact analysis of the transient response of linear antennas is possible only in some special cases. Infinitely long linear antennas are susceptible to exact analysis via Fourier (or Laplace) transform techniques or directly in the time domain. Fairly accurate expressions for the transient radiation from elementary linear radiators, e.g., the Hertzian dipole, may be obtained analytically by direct application of Fourier transform techniques to the known frequency domain results [4.47], if the input signal spectrum is such that the radiator may still be considered a Hertzian dipole. The reflectionless linear antenna is one of the few finite length antennas whose transient performance may be analyzed with a fair amount of accuracy. For a realistic linear antenna, the frequency domain current distribution is not generally known analytically over the frequency band of interest. Numerical techniques are the only available means for obtaining meaningful results and also the results obtained by the singularity expansion method are basically numerical in nature.

It would be quite useful to search for analytical solutions of the time domain integral equations for the current distributions on linear antennas excited by transient signals, even if the solutions are valid only over a certain period of time. The problem of radiated pulse shape when a linear antenna is excited by a rectangular pulse should be investigated. In fact, very little is known about the effects of the basic antenna parameters L and a on the radiated wave forms.

Although a certain amount of experimental investigation has been performed on the transient radiation and reception of linear antennas, we feel that experimental means have not been sufficiently exploited to gain an understanding of the transient characteristics of such antennas. Transient responses of linear antennas can be measured directly. Moreover, the measurements can be made in both the frequency and time domains. The availability of sampling oscilloscopes has revolutionized the time domain measurements. Currently available oscilloscopes have 50 picoseconds rise time (14 GHz bandwidth at $-3\,dB$) and sensitivities of several millivolts. Pulses as short as a few tenths of a nanosecond can be generated with modern semiconductor devices.

At present, numerical methods are the principal means available and are popular for studying many practical transient electromagnetic problems. However, on many occassions one finds it difficult to correlate the numerical results with the physical parameters of an antenna and also with the physical phenomenon itself. Carefully designed experiments will, we believe, clarify many aspects of the transient characteristics of linear antennas, which are presently only partly understood.

Acknowledgments

Part of this work was supported by the National Science Foundation under Grant GK 22898. The authors wish to express their sincere gratitude to Ms. CATHERINE RADER for her expert and patient typing of the manuscript.

References

4.1   C. MANNEBACK: AIEE J. **42**, 95 (1923)
4.2   S. A. SCHELKUNOFF: *Advanced Antenna Theory* (John Wiley and Sons Inc., New York 1952) pp. 102–109
4.3   C. POLK: Proc. IRE **4**, 1281 (1960)
4.4   B. R. MAYO: Proc. IRE **49**, 819 (1961)
4.5   B. R. MAYO, P. W. HOWELLS, W. B. ADAMS: Microwave J. **4**, 79 (1961)
4.6   F. T. TSENG, D. K. CHENG: Canad. J. Phys. **42**, 135 (1964). Also see D. K. CHENG, F. T. TSENG: IEEE Trans. AP-**12**, 492 (1964)
4.7   H. J. SCHMITT: Proc. IEE **107c**, 292 (1960)
4.8   H. J. SCHMITT: IEEE Trans. AP-**11**, 509 (1963)
4.9   H. J. SCHMITT, C. W. HARRISON, C. S. WILLIAMS: IEEE Trans. AP-**14**, 120 (1966)
4.10 R. W. P. KING, H. J. SCHMITT: IRE Trans. AP-**10**, 222 (1962)
4.11 C. W. HARRISON, C. S. WILLIAMS: IEEE Trans. AP-**13**, 236 (1965)
4.12 G. F. ROSS, R. H. T. BATES, G. HANLEY, K. ROBBINS, L. SUSMAN: Transient Behavior of Radiating Elements; Sperry-Rand Research Center, Final Report on AF30 (602)–4050, Sudbury, Mass. (1966)
4.13 T. T. WU: J. Math. Phys. **2**, 92 (1961)
4.14 S. P. MORGAN: J. Math. Phys. **3**, 564 (1962)
4.15 P. O. BRUNDELL, ERICSSON: Technics **16**, 137 (1960)
4.16 O. EINARSSON: Trans. Roy. Inst. Tech., No. 191 (1962)
4.17 R. W. LATHAM, K. S. H. LEE: Radio Sci. **5**, 715 (1970)
4.18 D. L. SENGUPTA, Y.-P. LIU: Radio Sci. **9**, 621 (1974)
4.19 C. W. HARRISON, R. W. P. KING: IEEE Trans. AP-**15**, 301 (1967)
4.20 A. M. ABO-ZENA, R. E. BEAM: IEEE Trans. AP-**19**, 129 (1971)
4.21 R. J. PALCIAUSKAS, R. E. BEAM: IEEE Trans. AP-**18**, 276 (1970)
4.22 Y.-P. LIU, D. L. SENGUPTA: IEEE Trans. AP-**22**, 212 (1974)
4.23 A. M. AUCKENTHALER, C. L. BENNETT: IEEE Trans. MTT-**19**, 892 (1971). Also see C. L. BENNETT, J. D. DeLORENZO, A. M. AUCKENTHALER: Integral Equation Approach to Wideband Inverse Scattering, Rome Air Development Center, Rome, New York, Rept. RADC-TR-78-177, Vol. 1 (1970)
4.24 A. J. POGGIO, E. K. MILLER: In *Computer Techniques for Electromagnetics*, ed. by R. MITTRA (Pergamon Press, Oxford, New York 1973), Chapt. 4, pp. 159–261
4.25 T. K. LIU, K. K. MEI: Radio Sci. **8**, 797 (1973)

4.26 E. P. SAYRE, R. HARRINGTON: Appl. Sci. Res. **26**, 413 (1972)

4.27 E. K. MILLER: Some Computational Aspects of Transient Electromagnetics, Lawrence Livermore Laboratory, University of California, Livermore, Rept. UCRL-51276 (1972)

4.28 F. M. TESCHE: IEEE Trans. AP-**21**, 53 (1973)

4.29 S. A. SCHELKUNOFF, H. T. FRIIS: *Antennas Theory and Practice* (John Wiley and Sons, Inc., New York 1952) pp. 361–394

4.30 G. SINCLAIR: Proc. IRE **38**, 148 (1950)

4.31 E. C. TITSCHMARSH: *Theory of Fourier Integrals*, 2nd ed. (Oxford at the Clarendon Press 1948) pp. 50–51

4.32 L. R. RABINER, C. M. RADER: *Digital Signal Processing* (IEEE Press, IEEE, Inc., New York 1972) pp. 223–35

4.33 Y.-P. LIU, D. L. SENGUPTA, C.-T. TAI: Electromagnetic Sensor Simulation Note 178, Air Force Weapons Laboratory (EL), Kirtland Air Force Base, New Mexico 87117, February 1973

4.34 R. F. HARRINGTON: *Field Computations by Moment Methods* (Macmillan, New York 1968) Chaps. 1 and 4

4.35 K. K. MEI: IEEE Trans. AP-**14**, 374 (1965)

4.36 T. T. WU, R. W. P. KING: IEEE Trans. AP-**12**, 369 (1965)

4.37 L. C. SHEN: IEEE Trans. AP-**15**, 606 (1967)

4.38 L. C. SHEN, T. T. WU: Radio Sci. **2** (New Series), 191 (1967)

4.39 C. L. BENNETT, W. L. WEEKS: IEEE Trans. AP-**18**, 627 (1969)

4.40 J. A. STRATTON: *Electromagnetic Theory* (McGraw-Hill Book Co., Inc., New York 1941) pp. 428–430

4.41 R. E. COLLIN, F. J. ZUCKER: *Antenna Theory*, Pt. I (McGraw-Hill Book Co., Inc., New York 1969) Chapt. 8

4.42 P. R. GARABEDIAN: *Partial Differential Equations* (John Wiley and Sons, Inc., New York 1964) Chapt. 13

4.43 P. D. LAX, R. D. RICHTMYER: Comm. Pure Appl. Math. **9**, 267 (1957)

4.44 R. D. RICHTMYER, K. W. MORTON: *Difference Methods for Initial-Value Problems*, 2nd ed. (Interscience Publishers, New York 1967) Chapt. 1 and 10

4.45 R. W. P. KING: *The Theory of Linear Antennas* (Harvard University Press, Cambridge, Mass. 1956) Chapt. 2

4.46 R. COURANT, D. HILBERT: *Methods of Mathematical Analysis*, Vol. 2 (Interscience Publishers, New York 1962) pp. 407–550

4.47 G. FRANCESCHETTI, C. PAPAS: IEEE Trans. Ap-**22**, 651 (1974)

5. A Pulsed Dipole in the Earth

J. A. FULLER* and J. R. WAIT**

With 13 Figures

Dispersion probably affects broad-band pulse propagation more ser-iously than any other electromagnetic phenomenon. Its influence has been discussed in general terms in Chapter 1, and has been illustrated on some idealized examples in Section 1.1. The presentation here deals with a realistic modeling of pulse propagation through the earth.

An electromagnetic (EM) pulse is progressively distorted as it propa-gates through the earth. The finite conductivity of the medium causes greater relative attenuation of the higher frequencies in the pulse spec-trum; a pronounced frequency dependence (dispersion) of the earth's constitutive parameters, particularly conductivity and permittivity, alters the spectrum; and spatial inhomogeneities such as ore bodies, the earth-air interface, and sedimentary layers introduce reflections and/or guided modes, which are delayed in time and are further distorted by the boundary interaction. Whether the source is buried for testing purposes, geophysical probing, or telecommunication, correct interpretation of the signal received at a distant point requires familiarity with propagation-induced distortion.

In this investigation analytical and computational techniques are applied to describe transient propagation from a short (Hertzian) dipole that is buried in poorly conducting rock. A dissipative half-space models the rock with conductivity and permittivity parameters that vary with frequency in a realistic manner. A vertical dipole orientation is assumed to permit placement in a drill hole. A similar arrangement is assumed for the receiver, and the vertical electric field component is examined. The assumed transient time dependence for the current moment of the transmitting dipole is an exponentially decaying step function. This system could be implemented by discharging a capacitor through a short dipole that has grounded ends.

* This work performed while at the University of Colorado, Boulder. Current address: Texas Instruments, Inc., P. O. Box 6015 M/S 327, Dallas, TX 75222, USA.

** Cooperative Institute for Research in Environmental Sciences (CIRES), University of Colorado, Boulder, CO 80302, USA.

The transient waveforms that describe the radiated fields are computed by Laplace transform procedures. General and exact formulations for the frequency domain (time-harmonic) fields of a buried dipole are well known [5.1]. However, in many interesting cases the Sommerfeld integrals (Hankel transforms) that occur must be evaluated by careful numerical procedures. Especially in these cases, a numerical evaluation of the Laplace inversion integral is required to obtain time domain (transient) solutions. Both the Hankel and Laplace transforms are improper (semi-infinite) integrals with oscillatory integrands, and computational efficiency is essential. A number of analytical and numerical techniques must be combined to achieve this.

The results presented here include waveforms that describe propagation in horizontal planes at great depths and near the earth's surface; a broad range of dispersive models for the constitutive parameters provides additional comparisons. It is apparent that typical dispersive parameters can significantly modify the pulse waveforms of both through-the-earth and ground wave components.

In the following discussion, Section 5.1 presents frequency domain models that describe the dispersive constitutive parameters; Section 5.2 discusses pulse propagation at great depths in the earth; and Section 5.3 concerns propagation between shallowly buried antennas, when the earth-air interface plays an important role.

5.1 Electrical Dispersion in Geological Media

For studies of steady-state, electromagnetic propagation, specified values of the constitutive parameters—conductivity σ, permittivity ε, and permeability μ—adequately characterize any isotropic medium. Certain minerals, particularly iron compounds, impart magnetic properties to rocks (e.g., magnetite, an iron oxide, and pyrrhotite, an iron sulfide). Excluding these minerals, the permeability of most rocks assumes the free space value, $\mu_0 = 4\pi \times 10^{-7}$ Henry/meter [Ref. 5.2, p. 44]. Therefore, electromagnetic energy dissipation and storage within these media are measured by σ and ε, respectively. As a function both of rock type and of electrical frequency, these parameters may vary by many orders of magnitude.

Rock conductivities are commonly observed to increase with frequency, while permittivities decrease (e.g. [5.3]). A two or three order of magnitude variation in these quantities over a 10^2 to 10^8 Hz frequency range is quite typical. Although the physical processes are not thoroughly understood, the presence of water greatly exaggerates this behavior.

Even within the high frequency (HF) range from 10^5 to 10^8 Hz, significant dispersion may be evident. There is some question as to the accuracy of core sample measurements, but *in situ* measurements also reveal HF dispersion (wet rock: GRUBB and WAIT [5.4]; dry rock: LYTLE [5.5]; permafrost: GRUBB [5.6], and LYTLE, et al. [5.7]).

At low frequencies, rock permittivities may assume surprisingly large values. The behavior is intimately associated with water content and temperature. At 100 Hz and ambient temperatures, the dried terrestrial samples examined by SAINT-AMANT and STRANGWAY [5.8] and the preserved (dry) lunar samples examined by CHUNG et al. [5.9] exhibited permittivities in the range $5\varepsilon_0$ to $60\varepsilon_0$; near 500 K the range is $10\varepsilon_0$ to $600\varepsilon_0$. The permittivity of free space is $\varepsilon_0 = 8.854 \times 10^{-12}$ Farad/meter.

For wet rocks representative data were reported by KELLER and LICASTRO [5.10] and by SCOTT et al. [5.3]. These papers and others indicate low frequency (100 Hz) permittivities as great as $10^6\varepsilon_0$. Drying typically reduces the numbers to less than $100\varepsilon_0$. While the precise values are questionable, due to partially recognized measurement difficulties, it seems quite likely that the true values for wet rocks may be greater than, say, $10^3\varepsilon_0$. Measurement difficulties are associated with finite sample size and with sample-electrode contacts, which are generally non-Ohmic [5.11,12].

The supporting evidence for large, low frequency permittivities, in addition to the large amount of sample data, is the induced polarization (IP) phenomenon observed by *in situ* measurement in geophysical exploration (e.g., [5.13, 14]). As a result of IP studies, three mechanisms are credited with producing large polarization effects in rocks: electrode polarization, membrane polarization, and interfacial or particle polarization. The first two are electrochemical effects [5.14,15] and the third is a generalized artificial dielectric effect [5.16,17].

When an electric field is applied to any material, free charges drift and bound charges are displaced until restoring forces balance the force of the applied field. This displacement of bound charge from its neutral position creates internal charge dipoles, which are directed oppositely to the applied electric field. Of course, the charge drift results in a conduction current density, $J_c = \sigma E$, and the induced dipoles produce an internal polarization (flux density), $P = (\varepsilon - \varepsilon_0)E$, where E is the electric field intensity. In steady-state situations, assuming a time factor $\exp(-i\omega t)$, the displacement current density is defined by $J_d = -i\omega\varepsilon E$. The loss tangent, $\tan \delta = \sigma/\omega\varepsilon$, measures the ratio of steady-state conduction to displacement currents within the medium. Measurements suggest that the loss tangent is often nearly constant over the frequency range 0.1 to 1000 Hz [18; Ref. 2, p. 44], indicating that $\sigma/\varepsilon \propto \omega$. If σ and ε are constant, then conduction currents predominate at low frequencies and displacement currents do at high frequencies. It is significant that this general behavior is not altered by frequency dependent, large, high frequency conductivities and large, low frequency permittivities. Measurement difficulties may be anticipated, simply because the large, rapidly varying parameters are associated with second-order conduction mechanisms.

Clearly, charge that appears to be elastically bound at low frequencies may at high frequencies behave much like unbound charge, because the field reverses so rapidly that neither type of charge moves far from its initial position. This argument applies to classical dielectrics, and, in fact, predicts that with increasing frequency the permittivity decreases and the conductivity increases [5.19]. Dielectrics typically have a dipole relaxation in the microwave range and have ionic and electronic cyclotron resonances at infrared and ultraviolet frequencies, respectively. However, the physical mechanisms usually associated with dielectrics are much too mild to explain the behavior of rocks.

The curious properties of rocks are somewhat related to the presence of nominally "free" charge carriers and to the inhomogeneous structure of rocks. Thermally liberated carriers may be expected to drift uniformly through a homogeneous solid, but in moving through a volume of rock there are numerous insulating boundaries, electrochemical layers, lattice defects, and impurity traps that may impede or block the charge flow. Resulting non-uniform charge distributions can lead to regions that are not charge-neutral and, therefore, to locally polarized regions.

Because of the interrelated roles of conduction and polarization mechanisms in rock, conductivity and permittivity cannot be independently defined [5.20–22]. At any frequency a well defined distinction does exist between conduction (in phase) and displacement (quadrature) currents, but the relative variations with frequency are constrained by the physical processes. In the present discussion the two variable constitutive parameters are described by a single "complex relative dielectric constant",

$$K(i\omega) = [\varepsilon_e(\omega) + i\sigma_e(\omega)/\omega]/\varepsilon_0 , \qquad (5.1)$$

where $\sigma_e(\omega)$ and $\varepsilon_e(\omega)$ are the (real-valued) effective conductivity and permittivity at the frequency ω.

It is assumed here that the rock is a linear medium. Of course, only with this assumption are the transient and steady state responses related by Laplace transforms. Although nonlinear behavior has been observed occasionally in laboratory rock samples [5.15,23], the low intensity of EM fields in the earth provides a reasonable justification for assuming in situ linearity [5.24].

The usual realizability criteria of linear-system theory impose constraints on the mathematical form of the complex dielectric constant. For example, the real and imaginary parts of a frequency domain representation must satisfy Hilbert transform, or Kramers-Krönig, relations (e.g. [5.25,26]). These conditions derive from the requirement that any realizable EM signal must be real-valued and causal.

A mathematical model for the complex dielectric constant is required which can be adjusted to fit experimental data in the frequency range from 10^2 to 10^8 Hz, which describes reasonable extrapolations to zero and infinite frequency limits, and which predicts physically realizable EM propagation. In geophysics, models are generally chosen without regard for their limiting behavior at high frequencies and, occasionally, without concern about realizability. For example, curve fitting techniques are sometimes applied to experimental data for conductivity and for permittivity, independently [5.27]. The Kramers-Krönig relations show that after the variation of one of these quantities is specified, the other is uniquely determined. However, realizability is not an issue in steady-state work, and the high frequency content of signals is not generally observed in geophysics. Certain mathematical liberties can be permitted in these cases.

For the present transient calculations, the earth's dielectric constant is modelled by a frequency-independent conductivity σ_0 and a number of Debye relaxation terms [5.28] as follows

$$K(s) = \frac{\sigma_0}{s\varepsilon_0} + \sum_{m=1}^{M} \left[a_m + \frac{b_m - a_m}{1 + s/\omega_m} \right], \tag{5.2}$$

where $s = -i\omega$ and the parameters are chosen to fit experimental data. Limiting values of the time-harmonic permittivity and conductivity are the following

$$\varepsilon_\infty/\varepsilon_0 = \varepsilon_e(\infty)/\varepsilon_0 = \sum_m a_m, \tag{5.3}$$

$$\varepsilon_e(0)/\varepsilon_0 = \sum_m b_m, \tag{5.4}$$

$$\sigma_\infty = \sigma_e(\infty) = \sigma_0 + \sum_m \omega_m \varepsilon_0 (b_m - a_m), \tag{5.5}$$

and

$$\sigma_e(0) = \sigma_0. \tag{5.6}$$

In a "relaxation" model an impulse electric field induces an exponentially decaying polarization. Each of the summation terms in (5.2) corresponds to a relaxation with a time constant, or relaxation time, ω_m^{-1}. The distributed relaxation times provide the flexibility needed to fit experimental data, and the general mathematical form satisfies the realizability criteria.

The quantitative behavior of six specific dispersion models, A through F, is described by curves of σ_e and ε_e in Fig. 5.1. These are the models employed for the pulse calculations. All six of these models include a static conductivity of $\sigma_0 = 10^{-5}$ mho/m; the remaining parameters, a_m, b_m, and $f_m = \omega_m/(2\pi)$, are given in Table 5.1[1]. Although parameters

Table 5.1. Parameters for dispersion models $\sigma_0 = 10^{-5}$[mho/m]

Model/m		1	2	3	4	5
A						
	a_m	15.0	3.0	1.0	0.7	0.3
	b_m	15.0	3.0	1.0	0.7	0.3
	f_m [Hz]	8.E+4	2.E+5	1.E+6	2.E+7	5.E+7
B						
	a_m	9.0	3.0	1.0	0.7	0.3
	b_m	15.0	3.0	1.0	0.7	0.3
	f_m [Hz]	8.E+4	2.E+5	1.E+6	2.E+7	5.E+7
C						
	a_m	9.0	0.0	0.0	0.0	0.0
	b_m	15.0	3.0	1.0	0.7	0.3
	f_m [Hz]	8.E+4	2.E+5	1.E+6	2.E+7	5.E+7
D						
	a_m	9.0	0.0	0.0	0.0	0.0
	b_m	45.0	3.0	1.0	0.7	0.3
	f_m [Hz]	8.E+4	2.E+5	1.E+6	2.E+7	5.E+7
E						
	a_m	9.0	0.0	0.0	0.0	0.0
	b_m	95.0	3.0	1.0	0.7	0.3
	f_m [Hz]	8.E+4	2.E+5	1.E+6	2.E+7	5.E+7
F						
	a_m	0.0	0.0	0.0	0.0	10.0
	b_m	9000.0	700.0	200.0	80.0	20.0
	f_m [Hz]	1.E+2	1.E+3	1.E+4	1.E+5	1.E+6

for five relaxation terms are included, the values are chosen so that model A has no relaxation, and model B has only one. The circled points in Fig. 5.1 represent experimentally determined values for permafrost [2.6], and the parameters for curve C are chosen to provide an approximate fit to this data. A very accurate fit may be obtained

---

[1] The notation 8.E+4 that is used in these tabulations implies 8.0×10^4.

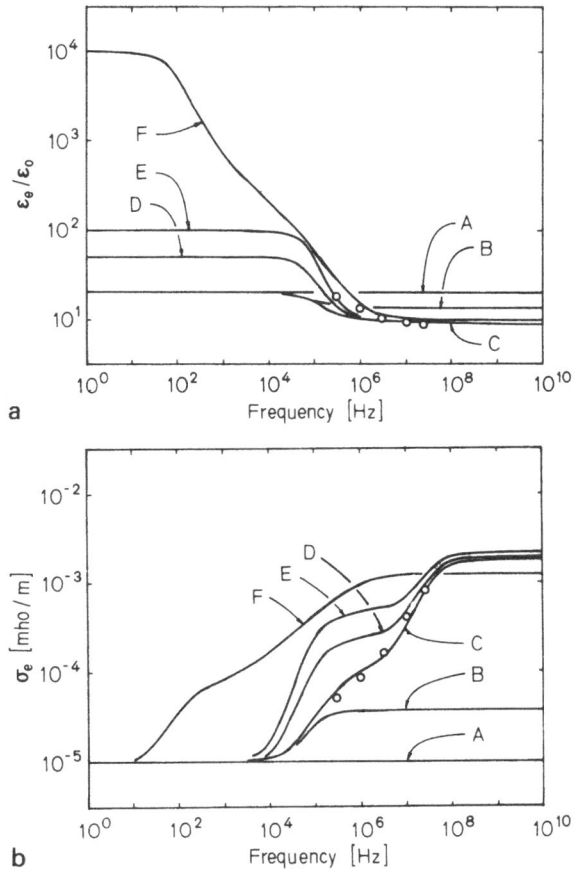

Fig. 5.1a and b. Frequency dependence of relaxation dispersion models. a) Relative permittivity, b) conductivity

for either ε_e or σ_e, and curve C represents a good compromise. With C as a reference, the sequence of models, A, B, and C, displays an increasing degree of high frequency dispersion, while the sequence, D, E, and F, displays increasing low frequency dispersion.

5.2 Dipole Radiation in a Homogeneous Earth

This section concerns propagation between dipoles placed in an infinite medium; the earth-air interface can be neglected when both the source and the observer are buried at sufficient depths. Fig. 5.2 describes the physical configuration in cylindrical coordinates (ϱ, ϕ, z). The dipole has a current moment $I(s)dl$ (Idl), is located at the origin of the coordinate system, and is aligned with the z-axis.

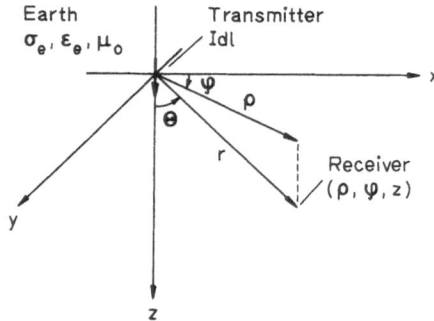

Fig. 5.2. Dipole antenna and receiver buried in an infinite, dispersive earth

The time-harmonic solution for the dipole fields is well-known[5.1]. The fields may be derived from a Hertz vector potential; only a z-component is required, and this is given by the solution of a wave equation,

$$(\nabla^2 - \gamma^2)\Pi(s) = -(\sigma_e + s\varepsilon_e)^{-1}J(s). \tag{5.7}$$

The complex frequency is $s = -i\omega$, and the complex propagation constant γ is given by

$$\begin{aligned}\gamma &= \gamma_0 K^{1/2}(s) \\ &= (s\mu_0\sigma_e + s^2\mu_0\varepsilon_e)^{1/2}, \end{aligned} \tag{5.8}$$

where $\gamma_0 = s/c_0 = s(\mu_0\varepsilon_0)^{1/2}$ is the free space propagation constant; $K(s)$ is the complex dielectric constant. The square roots are defined by principal values $(\mathrm{Re}\{\gamma\} \geq 0)$. The z-directed current density J is given by

$$J(s) = I(s)dl\,\delta(z)\delta(\varrho)(2\pi\varrho)^{-1}. \tag{5.9}$$

where $\delta(x)$ is the Dirac delta function. The solution for the Hertz potential is

$$\Pi(s) = \frac{I(s)dl\exp(-\gamma r)}{4\pi(\sigma_e + s\varepsilon_e)r} \tag{5.10}$$

with $r=(\varrho^2+z^2)^{1/2}$, and the resulting z-component $E(s)$ of the electric field is given by

$$E(s)=(\partial^2/\partial z^2 - \gamma^2)\Pi(s)$$

$$=\frac{I(s)dl\exp(-\gamma r)}{4\pi(\sigma_e+s\varepsilon_e)r^3}\cdot[(2-3\sin^2\theta)(1+\gamma r)-\sin^2\theta(\gamma r)^2]. \quad (5.11)$$

Pulse responses are evaluated by inverse Laplace transformation of (5.11). The transient excitation selected in the present case is an exponentially decaying step function. The current moment is described by

$$I(t)dl=I_0dle^{-\alpha t}U(t), \quad (5.12)$$

where $U(t)$ is the unit step function, and the Laplace transform is

$$I(s)dl=I_0dl/(s+\alpha). \quad (5.13)$$

The pulse response is computed by the transform inversion integral,

$$E(t)=\frac{1}{2\pi i}\int_{a-i\infty}^{a+i\infty}E(s)e^{st}ds, \quad (5.14)$$

where the real number a is chosen so that no singularities of E lie to the right of the integration path.

For a non-conducting earth with constant permittivity ε, the Laplace inversion may be performed easily by analytical procedures. The step function response $(\alpha=0)$ in the horizontal plane $(\theta=\pi/2)$ is simply

$$E(t)=\frac{I_0dl}{4\pi\varepsilon\varrho^3}[tU(t-t_0)+t_0^2\delta(t-t_0)], \quad (5.15)$$

where the arrival time is $t_0=\varrho/c$ with a phase velocity $c=(\mu_0\varepsilon)^{-1/2}$. However, with finite conductivity various approximations are required to obtain analytical results [5.29–31]. Series expansions adequately describe the early portion of the pulse, and solutions that neglect displacement current are useful for the late portion; but a numerical evaluation of (5.14) is needed to determine the complete pulse, especially when the effects of dispersion are of concern.

The transform is computed by an adaptive application of a method described by LUKE [5.32]. This method evaluates the basic Fourier integral

$$F(t)=\int_{\omega_1}^{\omega_2}F(\omega)e^{-i\omega t}d\omega \quad (5.16)$$

by interpolating $F(\omega)$ with a polynomial and then utilizing well-known integral formulas. This is an extension of Filon's method to polynomials of degree higher than two. However, Luke's algorithm is formulated in terms of the central differences of $F(\omega)$. The central differences provide a convenient indication of the accuracy of the interpolation. In this application the interval $[\omega_1, \omega_2]$ is adaptively halved until the fourth central difference indicates that a five-point interpolation is adequate. This adaptive approach is employed in an effort to minimize the number of points at which $F(\omega)$ must be evaluated. For the same reason, the same set of $F(\omega)$ values is used for all values of time t. A convergence accelerating $(G-)$ transformation is applied to estimate the improper integral, as the upper limit of integration is increased [5.33]. When the truncated integration agrees with the G-transform estimate, the estimate is presumed to be a valid answer.

At high frequencies $(s \to \pm i\infty)$ the real part of γ approaches a constant; therefore, the inversion integral is not rapidly convergent. The situation is improved by analytically inverting the high frequency portion, because then the high frequency asymptote can be subtracted from the integrand. Employing the high frequency limiting values in (5.3) and (5.5), the complex dielectric constant is approximated by

$$K(s) \cong (\varepsilon_\infty/\varepsilon_0)(1 + \beta/s),\tag{5.17}$$

where $\beta = \sigma_\infty/\varepsilon_\infty$; it follows that

$$\gamma \cong (s/c)(1 + \beta/s)^{1/2} \sim (s/c)[1 + (\beta/s)/2 - (\beta/s)^2/8 + (\beta/s)^3/16].\tag{5.18}$$

Substituting into (5.11) and retaining numerator terms of constant and higher orders in s yields an approximate high frequency representation $E_\alpha(s)$,

$$E_\alpha(s) = \frac{I_0 dl}{4\pi\varepsilon_\infty r^3} \cdot \frac{C_2 s^2 + C_1 s + C_0}{(s+\alpha)(s+\beta)},\tag{5.19}$$

where

$$C_0 = \left[1 + \frac{\beta t_0}{2} + \frac{(\beta t_0)^2}{8}\right](2 - 3\sin^2\theta)$$
$$- \left[\frac{(\beta t_0)^3}{16} + \frac{(\beta t_0)^4}{128}\right]\sin^2\theta,\tag{5.20}$$

$$C_1 = t_0\left[2 - 3\sin^2\theta - \left(\beta t_0 + \frac{(\beta t_0)^2}{8}\right)\sin^2\theta\right],\tag{5.21}$$

and

$$C_2 = -t_0^2 \sin^2 \theta. \tag{5.22}$$

The elementary inverse of (5.19) is

$$E_\alpha(t) = \frac{I_0 dl \exp(-\beta t_0/2)}{4\pi\varepsilon_\infty r^3} \left\{ C_2 \delta(t-t_0) \right.$$
$$+ \left[(C_2\alpha^2 - C_1\alpha + C_0)e^{-\alpha(t-t_0)} \right. \tag{5.23}$$
$$\left. - (C_2\beta^2 - C_1\beta + C_0)e^{-\beta(t-t_0)} \right] \frac{u(t-t_0)}{\beta - \alpha} \left. \right\}.$$

The approximation $E_\alpha(t)$ is valid for times near the arrival time t_0 and specifies the initial impulse and step discontinuities,

$$E_\delta = \frac{I_0 dl \exp(-\beta t_0/2)}{4\pi\varepsilon_\infty r^3} C_2 \tag{5.24}$$

and

$$E_U = \frac{I_0 dl \exp(-\beta t_0/2)}{4\pi\varepsilon_\infty r^3} [C_1 - C_2(\alpha+\beta)]. \tag{5.25}$$

When $\alpha = 0$, these quantities and, also, the initial slope agree with previously reported results (e.g. [5.31]).

Results of the numerical integration are shown by the solid curves in Figs. 5.3 and 5.4. The transmitter and receiver are placed in the horizontal plane ($z=0$); the radial separation is $\varrho = 1000$ m in Fig. 5.3 and $\varrho = 100$ m in Fig. 5.4. The curves in each figure are designated by letters that correspond to the dielectric models of Fig. 5.1. The arrival times t_0 and the amplitudes of the initial impulse and step discontinuities, E_δ and E_U, are provided in Table 5.2. The exciting pulse decays with a one-tenth second time constant ($\alpha = 10\,\text{s}^{-1}$).

Although $E_\alpha(t)$ is developed as a short time approximation, it appears to be useful at longer times, provided that the low frequency parameters $\varepsilon_e(0)$ and $\sigma_e(0)$ are employed in (5.23). In Table 5.3 the numerically computed pulse response is compared with $E_\alpha(t)$. In these cases the conductivities and permittivities are constant and assume the low frequency (LF) limiting values of the indicated dielectric models (e.g., A, B, and C limit to $\sigma_e(0) = 10^{-5}$ mho/m, $\varepsilon_e(0) = 20\varepsilon_0$; and F limits to $\sigma = 10^{-5}$ mho/m, $\varepsilon = 10^4\varepsilon_0$). In the table the column headed $E(t)/I_0 dl$

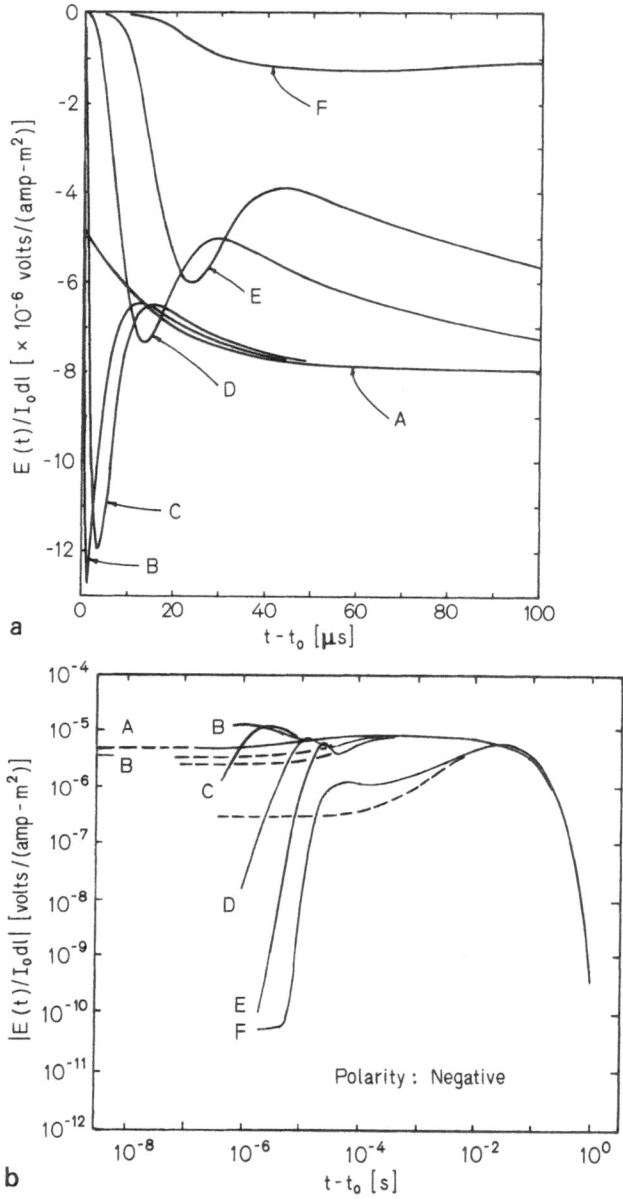

Fig. 5.3. a) Pulse response in homogeneous, dispersive earth (linear scales); equatorial plane of the dipole ($\varrho = 1000$ m, $z = 0$). In this time range the source waveform is essentially a step function. b) Pulse response in homogeneous, dispersive earth (log scales); equatorial plane of the dipole ($\varrho = 1000$ m, $z = 0$). Dashed curves represent approximation $E_\alpha(t)/I_0 dl$ with low frequency limiting parameters $\sigma_e(0)$, $\varepsilon_e(0)$. "Polarity" refers to the arithmetic sign of the function

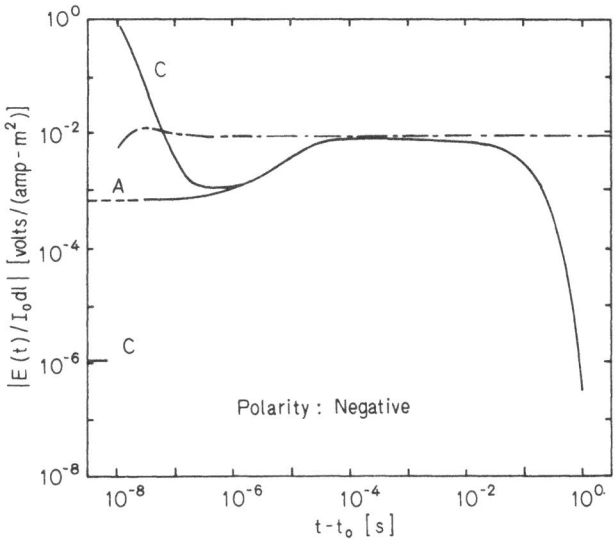

Fig. 5.4. Pulse response in homogeneous, dispersive earth; equatorial plane of the dipole $(\varrho = 100$ m, $z = 0)$. Dashed curve represents approximation $E_\alpha(t)$ with the low frequency parameters $\sigma_e(0)$ and $\varepsilon_e(0)$. Broken line represents diffusion approximation to *step* response, $E_d(t)$

Table 5.2. Initial parameters for pulses in homogeneous earth
$\varrho = 10^3$ m; $z = 0$

Dielectric model	t_0 [μs]	$E_\delta/I_0 dl$ [volt-s)/(amp-m²)]	$E_U/I_0 dl$ [volt/(amp-m²)]
A	14.92	-6.563 E-11	-4.789 E-06
B	12.48	-1.576 E-11	-3.418 E-06
C	10.01	-3.787 E-58	-2.256 E-49
D	10.01	-8.660 E-62	-5.981 E-53
E	10.01	-7.403 E-68	-6.400 E-59
F	10.55	-4.521 E-42	-1.116 E-33

is the direct result of the truncated numerical integration, and the column headed $G[E(t)/I_0 dl]$ results from application of the G-transform. As expected, $G[E(t)/I_0 dl]$ and $E_\alpha(t)/I_0 dl$ compare well at early times, but they agree almost equally well at later times.

The dashed curves in Fig. 5.3b represent pulse responses when the *constant* parameters $\sigma_e(0)$ and $\varepsilon_e(0)$ describe the medium. These media may be identified as "non-dispersive", low frequency equivalents of the dispersive models in Fig. 5.1. In comparison with the solid curves,

Table 5.3. Comparison of numerical and approximate pulse responses using low frequency dielectric parameters

Horizontal Plane: $\varrho = 1000$ m; $z = 0$

$t - t_0$	$E(t)/I_0 dl$	$G[E(t)/I_0 dl]$	$E_\alpha(t)/I_0 dl$
	Models A, B, C: $\sigma = 10^{-5}$ mho/m; $\varepsilon = 20\varepsilon_0$ (LF limit)		
1.E-6	−4.90 E-06	−4.98 E-06	−4.97 E-06
2.E-6	−5.11 E-06	−5.14 E-06	−5.14 E-06
5.E-6	−5.59 E-06	−5.60 E-06	−5.60 E-06
1.E-5	−6.20 E-06	−6.21 E-06	−6.22 E-06
2.E-5	−7.01 E-06	−7.01 E-06	−7.03 E-06
5.E-5	−7.84 E-06	−7.84 E-06	−7.90 E-06
1.E-4	−7.98 E-06	−7.98 E-06	−8.08 E-06
2.E-4	−7.96 E-06	−7.96 E-06	−8.08 E-06
5.E-4	−7.92 E-06	−7.92 E-06	−8.06 E-06
1.E-3	−7.88 E-06	−7.88 E-06	−8.02 E-06
1.E-2	−7.20 E-06	−7.20 E-06	−7.33 E-06
1.E-1	−2.93 E-06	−2.93 E-06	−2.98 E-06
1.E-0	−3.58 E-10	−3.58 E-10	−3.68 E-10
	Model F: $\sigma = 10^{-5}$ mho/m; $\varepsilon = 10^4 \varepsilon_0$ (LF limit)		
1.E-6	2.09 E-06	−2.94 E-07	−2.94 E-07
2.E-6	8.92 E-07	−2.96 E-07	−2.95 E-07
5.E-6	1.79 E-07	−2.96 E-07	−2.98 E-07
1.E-5	−6.46 E-08	−3.02 E-07	−3.02 E-07
2.E-5	−1.92 E-07	−3.11 E-07	−3.11 E-07
5.E-5	−2.89 E-07	−3.36 E-07	−3.36 E-07
1.E-4	−3.55 E-07	−3.79 E-07	−3.79 E-07
2.E-4	−4.52 E-07	−4.64 E-07	−4.64 E-07
5.E-4	−7.07 E-07	−7.12 E-07	−7.12 E-07
1.E-3	−1.10 E-06	−1.10 E-06	−1.10 E-06
1.E-2	−5.15 E-06	−5.16 E-06	−5.16 E-06
1.E-1	−3.20 E-06	−3.20 E-06	−3.20 E-06
1.E-0	−3.88 E-10	−3.91 E-10	−3.95 E-10

these waveforms show a larger step discontinuity at the time of the pulse arrival. These "non-dispersive" media have a conductivity of 10^{-5} mho/m at all times, while the dispersive models have effective values of approximately 10^{-3} mho/m at short times. The high initial conductivity apparently attenuates and delays the early response in the dispersive media.

The broken curve in Fig. 5.4 shows a diffusion approximation to the step function response $(\alpha = 0)$. As discussed by WAIT and SPIES [5.31], the approximation is valid for $\sigma/\varepsilon \ll t$. For $z = 0$, it is given by

$$E_d(t) = \frac{I_0 dl}{4\pi\sigma r^3}\left[\text{erfc}(x) + (x + 2x^3)2\pi^{-1/2}e^{-x^2}\right]U(t) \tag{5.26}$$

where erfc(x) is the complementary error function and $x = [\sigma \mu_0 r^2/(4t)]^{1/2}$. The one-tenth second time constant of the decaying pulse $E(t)$ is sufficiently long that $E(t)$ accurately describes the early, transitory portion of the step response.

5.3 Dipole Radiation in a Half-Space

This section is concerned with dipole radiation in a semi-infinite earth. As before, the antennas are vertical electric dipoles (VED). Emphasis is placed on a configuration in which both the transmitter and receiver are buried near the earth-air interface. Fig. 5.5 shows the cylindrical coordinate system; the $z=0$ plane designates the earth-air interface, and the z-axis is directed into the earth. The transmitting dipole is buried at a depth h and is collinear with the z-axis.

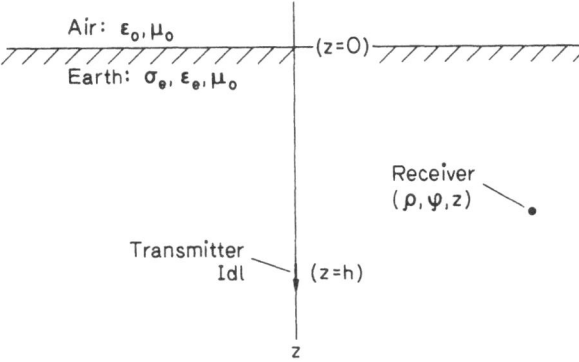

Fig. 5.5. Dipole antenna and receiver buried in a dispersive half-space

The well-known Sommerfeld formulation for the time-harmonic solution of the half-space problem is followed here [5.1]. The Hertz vector potential again requires only a z-component. In the earth the component $\Pi(s)$ satisfies (5.7), and in the air the component satisfies a homogeneous wave equation,

$$(\nabla^2 - \gamma_0^2) \Pi_0(s) = 0. \tag{5.27}$$

The boundary conditions require that

$$\frac{\partial \Pi(s)}{\partial z} = \frac{\partial \Pi_0(s)}{\partial z}\bigg|_{z=0} \tag{5.28}$$

and

$$\gamma^2 \Pi(s) = \gamma_0^2 \Pi_0(s)|_{z=0} . \tag{5.29}$$

The potential in the earth is conveniently represented by primary and secondary components

$$\Pi(s) = \Pi^{(p)}(s) + \Pi^{(s)}(s) . \tag{5.30}$$

The primary contribution omits interface effects and follows directly from Section 5.2[2],

$$\Pi^{(p)}(s) = \frac{I(s) dl \exp(-\gamma r_1)}{4\pi(\sigma_e + s\varepsilon_e) r_1}$$

$$= \frac{I(s) dl}{4\pi\varepsilon_0 s K} \int_0^\infty J_0(\lambda \varrho) \exp(-w|h-z|)(\lambda/w) d\lambda \tag{5.31}$$

where $r_1 = (\varrho^2 + (h-z)^2)^{1/2}$, and the remaining secondary contribution is

$$\Pi^{(s)}(s) = \frac{I(s) dl}{4\pi\varepsilon_0 s K} \int_0^\infty R(\lambda) J_0(\lambda \varrho) \exp(-w(h+z))(\lambda/w) d\lambda . \tag{5.32}$$

In these formulas J_0 is a Bessel function, and

$$R(\lambda) = \frac{w - K w_0}{w + K w_0} \tag{5.33}$$

with

$$w = (\lambda^2 + \gamma^2)^{1/2} \tag{5.34}$$

and

$$w_0 = (\lambda^2 + \gamma_0^2)^{1/2} ; \tag{5.35}$$

---

[2] In the following, λ is an integration variable and should not be confused with the same symbol used for wavelength elsewhere in the book.

the square roots are chosen so that $\text{Re}\{w\} \geq 0$ and $\text{Re}\{w_0\} \geq 0$ when λ is real, with $W = \gamma$ and $w_0 = \gamma_0$ when $\lambda = 0$.

Singularities of the integrand in (5.32) are the branch points at $\pm\lambda_0$ and $\pm\lambda_g$, where

$$\lambda_0 = i\gamma_0 , \tag{5.36}$$

$$\lambda_g = i\gamma , \tag{5.37}$$

and poles at $\pm\lambda_p$, where

$$\lambda_p = i\gamma_0 \left(\frac{K}{K+1}\right)^{1/2} , \tag{5.38}$$

which occur when $w + Kw_0 = 0$. With the vertical branch cuts illustrated in Fig. 5.6, the poles of $R(\lambda)$ lie on hidden (lower) sheets, and a zero $(w - Kw_0 = 0)$ lies on the visible sheet at $\lambda = \lambda_p$. The integration path is indented into the fourth quadrant to avoid the branch point at λ_0.

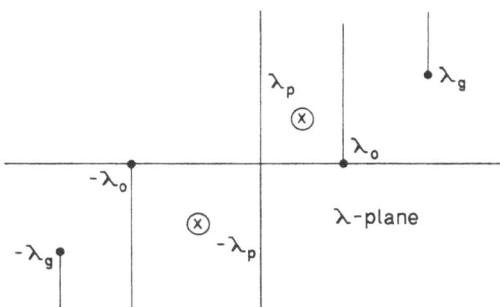

Fig. 5.6. Singularities in the integrand for the secondary potential

With the identification $\lambda = i\gamma \sin\theta$, the formulas (5.31) and (5.32) may be interpreted as integrals over a spatial spectrum of plane waves that propagate toward the interface with complex incidence angles θ. The factor $R(\lambda)$ in the secondary potential is a Fresnel reflection coefficient that accounts for the interface. The points $\lambda = \lambda_p$ and $\lambda = \lambda_0$ correspond with the Brewster and critical angles that are associated with plane wave refraction at a dielectric interface.

In what follows it is convenient to change the integration variable along the real λ-axis to $X=\lambda\varrho$ and to introduce the dimensionless quantities $D_0=|h-z|/\varrho$ and $D=(h+z)/\varrho$. The complex λ-plane is replaced by a Z-plane with $Z=X+iY$. The branch points in the Z-plane are $\pm Z_0=\pm\lambda_0\varrho$ and $\pm Z_g=\pm\lambda_g\varrho$; the pole location is $\pm Z_p=\pm\lambda_p\varrho$. The functions w_0 and w are replaced by $W_0=(X^2-Z_0^2)^{1/2}$ and $W=(X^2-Z_g^2)^{1/2}$; the prescription for the choice of roots is unchanged, and, except for the scale factor ϱ, Fig. 5.6 is unchanged.

The electric field is then given by

$$E(s) = \frac{I(s)\,dl}{4\pi\varepsilon_0\,s\,K\,\varrho^3}\left[P_0(D_0)+P(D)\right], \tag{5.39}$$

where P_0 describes the spherical wave behavior

$$
\begin{aligned}
P_0(D_0) &= \left[\frac{\partial^2}{\partial D_0^2} + Z_g^2\right]\frac{\exp\left[iZ_g(1+D_0^2)^{1/2}\right]}{(1+D_0^2)^{1/2}}\\
&= \int_0^\infty J_0(X)\exp(-WD_0)\frac{X^3\,dX}{W},
\end{aligned}
\tag{5.40}
$$

and P provides the secondary correction

$$P(D)=\int_0^\infty R(X)J_0(X)\exp(-WD)\frac{X^3\,dX}{W}, \tag{5.41}$$

with $R(X)=(W-KW_0)/(W+KW_0)$.

For shallow burial depths, a further expansion of the secondary contribution conveniently exhibits "image" and "ground wave" contributions. This representation follows from

$$P(D)=2P_g(D)-P_0(D), \tag{5.42}$$

where $P_0(D)$ is a spherical wave originating from the image point $z=-h$, and

$$P_g(D)=\int_0^\infty J_0(X)\exp(-WD)\frac{X^3\,dX}{W+KW_0}. \tag{5.43}$$

The primary and image components are easily described in closed form by appropriate substitution for r and θ in (5.11). For the primary spherical wave, use $r_1 = [\varrho^2 + (h-z)^2]^{1/2}$ and $\theta_p = \arctan(\varrho/|h-z|)$; for the secondary component, use $r_2 = [\varrho^2 + (h+z)^2]^{1/2}$ and $\theta_s = \arctan[\varrho/(h+z)]$.

The integrals $P(D)$ or $P_g(D)$ must be evaluated by numerical techniques. Except when D is small, a rather straightforward computation of $P(D)$ is possible [5.34]. For example, a Gaussian quadrature rule may be applied between the zeros of the Bessel function; special attention is required only when the integration contour passes near the pole λ_p and when the integrand oscillates rapidly. The former problem may be handled by a judicious choice of subintervals and/or by analytically substracting the singularity. The latter problem occurs primarily when $D \to 0$ (shallow burials) and may be handled by deforming the integration contour so as to surround the branch cuts in the upper half plane of Fig. 5.6. This deformation is readily accomplished after changing the integration interval to $[-\infty, \infty]$ and replaicing $J_0(x)$ with $(1/2)H_0^{(1)}(x)$ [5.35].

For cases when D is small, the representation (5.42) is utilized, and $P_g(D)$ is given by the sum of the branch line integrals P_1, around Z_g, and P_2, around Z_0,

$$P_g(D) = P_1(D) + P_2(D) . \tag{5.44}$$

The vertical branch cut which originates at Z_g is described by $Z = Z_g + iq^2$, where q^2 is real and positive. The contour integration around this cut is

$$P_1(D) = i \int_0^\infty H_0^{(1)}(Z_g + iq^2)(Z_g + iq^2)^3 \left[\frac{e^{-DW}}{W + K W_0} + \frac{e^{DW}}{W - K W_0} \right] q \, dq , \tag{5.45}$$

where

$$W = (i 2 Z_g q^2 - q^4)^{1/2} \tag{5.46}$$

and

$$W_0 = (Z_g^2 - Z_0^2 + i 2 Z_g q^2 - q^4)^{1/2} ; \tag{5.47}$$

principal values of the square roots are implied. Similarly, the branch cut that originates at Z_0 is described by $Z = Z_0 + iq^2$. The contour integration around this second cut is

$$P_2(D) = i \int_0^\infty H_0^{(1)}(Z_0 + iq^2) e^{-DV} (Z_0 + iq^2)^3$$

$$\cdot [(V + K W_0)^{-1} - (V - K W_0)^{-1}] q \, dq \qquad (5.48)$$

$$= -i 2 K \int_0^\infty H_0^{(1)}(Z_0 + iq^2) e^{-DV} \frac{W_0(Z_0 + iq^2)^3}{W^2 - K^2 W_0^2} q \, dq,$$

where

$$W_0 = (i 2 Z_0 q^2 - q^4)^{1/2}, \qquad (5.49)$$

$$W = (Z_0^2 - Z_g^2 + i 2 Z_0 q^2 - q^4)^{1/2}, \qquad (5.50)$$

with principal roots implied, and

$$V = W \, \mathrm{sgn}(-q^2 - \tfrac{1}{2} Z_0 \, \mathrm{Im}\{K\}) \qquad (5.51)$$

with $\mathrm{sgn}(x) = x/|x|$. With this formulation, convergence is provided by the Hankel function $H_0^{(1)}$, rather than by the exponential, and small values of D present no problem.

The primary difficulties associated with the P_1 and P_2 integrals are (i) the evaluation of Hankel functions with complex arguments, (ii) the near proximity of the pole Z_p to the integration contour of P_2 when $1 \ll |K|$, and (iii) oscillation of the exponential functions in the integrands when D is not small.

With small arguments, the basic power series expansions for $J_0(Z)$ and $Y_0(Z)$ yield accurate evaluations of the Hankel functions [5.36] With large arguments, Hankel's asymptotic expansion could be used [Ref. 37, p. 198]. However, a much better procedure is to use the integral representation from which that saddle point expansion is derived [Ref. 37, p. 168]. The integral representation may be easily and accurately evaluated by a generalized Gaussian-Laguerre quadrature. The required weights and abscissas for integrals of the form

$$\int_0^\infty f(v) v^{-1/2} \exp(-v) dv$$

are tabulated by SHAO et al. [5.38]. In the present application, a sixteen-point quadrature formula provides sufficient accuracy, when the argument of the Hankel function is greater than unity. Therefore, the integral quadrature is a suitable complement to the power series computation.

In the integrand of P_2, the point $q = 0$ corresponds to the branch point (and saddle point) Z_0, and a major contribution to P_2 may be expected in the region near that point. When $1 \ll |K|$, the pole Z_p approaches Z_0; the singularity is displaced from the contour of integration, but its presence cannot be ignored. However, the same techniques of pole subtraction that are frequently employed to facilitate approximate saddle point integration [5.39,40] may be applied directly to the present numerical integral [5.41].

Because this formulation in terms of branch line integrals is specifically intended for case with $D \ll 1$, the difficulty with oscillations of the exponential functions is not serious. Therefore, the subintervals for numerical integration are selected to provide constant increments in the variable q^2. To extend the range of D, it would be necessary to choose integration subintervals according to the period of oscillation.

In the low frequency limit $s \to 0$, a static approximation of P_g may be easily obtained. After letting $\gamma_0 = -iZ_0/\varrho \to 0$ and $\gamma = -iZ_g/\varrho \to 0$, the result is

$$P_g(D) \cong \frac{-1}{K(1+D^2)^{3/2}} \left(1 - \frac{3D}{1+D^2} \right). \tag{5.52}$$

For the particular case of $D=0$, the static approximation to the field is simply

$$E_0(s) = \frac{I(s)\,dl\,2\,P_g(0)}{4\pi\varepsilon_0\,sK\varrho^3} \cong \frac{-I\,dl\,s\varepsilon_0}{2\pi\sigma_0^2\varrho^3}. \tag{5.53}$$

A high frequency approximation to the ground wave field can also be derived rather easily by saddle point techniques. When the buried transmitter and receiver are both near the interface, the branch line contour integrals readily yield the lateral wave [5.42],

$$E_{lat}(s) = \frac{-I(s)\,dl}{2\pi(K-1)\varrho^2} \left(\frac{\mu_0}{\varepsilon_0} \right)^{1/2} \exp\{ -\gamma_0[\varrho + (h+z)(K-1)^{1/2}] \}. \tag{5.54}$$

This field displays the characteristic ϱ^{-2} dependence when $(h+z)/\varrho \ll 1$. The lateral wave represents a portion of the transmitted energy that is critically refracted at the earth-air interface. The wave spreads radially over the interface and continually "leaks" energy back into the earth. Because propagation velocities are greater in the air than in rock, the lateral wave is identified in pulse experiments by its early arrival time (see also Subsect. 1.5.3).

Table 5.4 provides a comparison between numerically integrated field values $E(s)/I(s)dl$ and the approximations $E_0(s)/I(s)dl$ and

Table 5.4. Comparison of exact and approximate fields
$\varrho = 10^3$ m, $\sigma_e = 10^{-3}$ mho/m
$h = z = 0^+$ m, $\varepsilon_e/\varepsilon_0 = 10$

f [Hz]	$E(s)/I(s)dl$	$E_{lat}(s)/I(s)dl$
1.E-01	8.854 192 E-16 $\angle +$ 90.00	
1.E-00	8.854 192 E-15 $\angle +$ 90.00	
1.E+01	8.854 192 E-14 $\angle +$ 90.00	
1.E+02	8.854 161 E-13 $\angle +$ 90.01	
1.E+03	8.852 005 E-12 $\angle +$ 90.06	
1.E+04	8.666 053 E-11 $\angle +$ 90.25	3.335 599 E-08 $\angle +$ 102.30
2.E+04	1.639 969 E-10 $\angle +$ 88.08	
5.E+04	4.984 998 E-10 $\angle +$ 58.11	
1.E+05	3.799 489 E-09 $\angle +$ 75.30	3.331 467 E-07 $\angle -$ 147.05
2.E+05	3.137 356 E-08 $\angle -$ 167.43	
5.E+05	4.127 951 E-07 $\angle -$ 135.86	
1.E+06	1.909 706 E-06 $\angle +$ 154.78	2.982 661 E-06 $\angle -$ 122.57
1.E+07	6.532 904 E-06 $\angle -$ 70.95	6.533 030 E-06 $\angle -$ 62.99
1.E+08	6.660 716 E-06 $\angle +$ 21.12	6.660 727 E-06 $\angle +$ 21.91
1.E+09	6.662 042 E-06 $\angle +$ 50.38	6.662 042 E-06 $\angle +$ 50.46
1.E+10	6.662 056 E-06 $\angle -$ 34.27	6.662 055 E-06 $\angle -$ 34.27
1.E+11	6.662 065 E-06 $\angle -$ 162.56	6.662 056 E-06 $\angle -$ 162.55

$E_{lat}(s)/I(s)dl$. With the parameters $\varrho = 10^3$ m, $z = h = 0$, and $\sigma_e = 10^{-3}$ mho/m, the low frequency approximation is given by

$$E_0(s)/I(s)dl = if(8.854 \times 10^{-15}) \tag{5.55}$$

where f is the frequency in Hz. Computed values for $E_{lat}(s)/I(s)dl$ are tabulated. It is apparent that $E(s)$ agrees well with $E_0(s)$ and $E_{lat}(s)$ at the low and high ends of the frequency range, respectively.

Pulse responses for the half-space problem are computed by the same numerical inverse Laplace transformation that is discussed in Section 5.2. As before, the exciting current moment is described by (5.13) with $\alpha = 10\,\text{s}^{-1}$, and the dispersive dielectric models of Fig. 5.1 are employed.

Pulse responses in three horizontal planes, $h = z = \infty$, 10, and 0, are shown in Fig. 5.7. The lateral separation between the transmitter and receiver is $\varrho = 100\,\text{m}$, and the permafrost model, C, describes the rock medium. At sufficient depths the interface has no effect, and the curve for $h = z = \infty$ is reproduced from Fig. 5.4, the result for a homogeneous medium. The pulse response at the 10 m depth is based on a numerical evaluation of the interface correction integral $P(D)$ in (5.41), and the response immediately below the interface (0 m) is computed using the ground wave integral $P_g(D)$ in (5.43). The initial step discontinui-

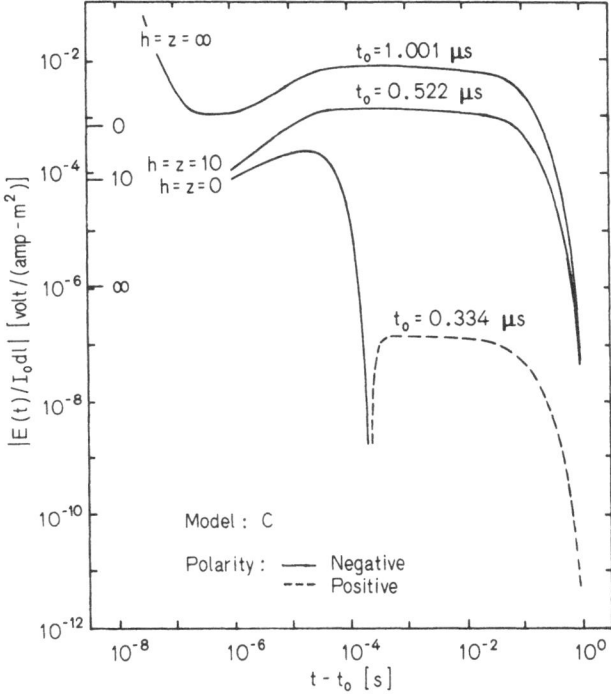

Fig. 5.7. Pulse response in a dispersive half-space ($\varrho = 100$ m)

ties for these pulses are indicated by the labeled tick marks on the ordinate.

The ground wave pulse, $h = z = 0$ in Fig. 5.7, is replotted on a linear scale in Fig. 5.8a. The initial sharp, but finite, spike and the slowly decaying tail are more evident in this figure. The same curve is included with two approximate representations in Fig. 5.8b. The approximation $E_0(t)$ is obtained by simply inverting the frequency domain representation $E_0(s)$ in (5.53). Substituting (5.13) for the current moment yields

$$E_0(t)/I_0 dl = \frac{-\varepsilon_0}{2\pi\sigma_0^2\varrho^3} L^{-1}\left\{\frac{s}{s+\alpha}\right\}$$

$$= \frac{\varepsilon_0}{2\pi\sigma^2\varrho^3} e^{-\alpha t} U(t),$$

(5.56)

where $L^{-1}\{\ \}$ represents the inverse transform (5.14). The approximation $E_1(t)$ is an extension of a transient solution described by VAN DER

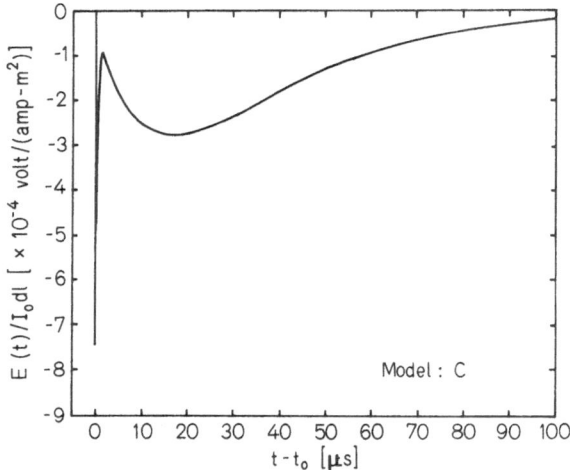

Fig. 5.8a. Pulse response in a dispersive half-space ($\varrho = 100$ m; $h=z=0$; $t_0=0.334$ μs)

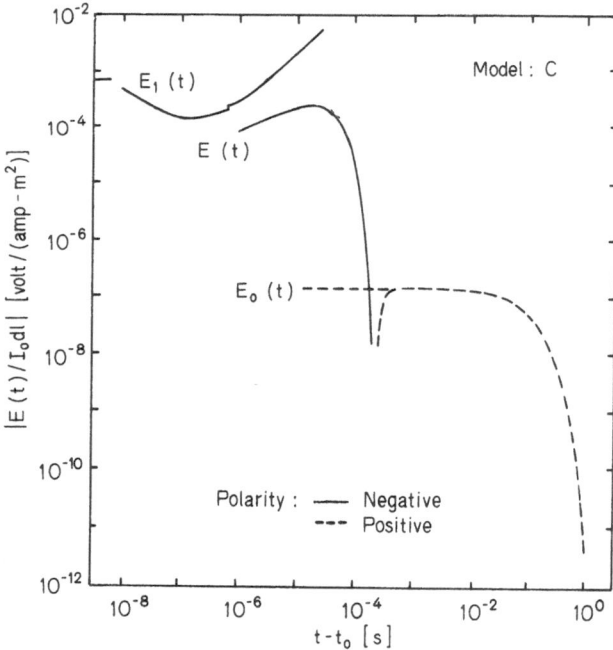

Fig. 5.8b. Pulse response in a dispersive half-space, with approximations E_1 and E_0
($\varrho = 100$ m; $h=z=0$; $t_0=0.334$ μs)

POL [5.43]. He examined the *impulse* response when the source (VED) and observer are both on the surface of a lossless dielectric half-space, and he obtained a closed form solution that is exact. The extension $E_1(t)$ describes the *step* response at the surface of a dielectric half-space. It is exact only for a lossless dielectric; but it is a useful approximation to the very early response, when the lateral wave arrives via a path through air.

For points on the interface, VAN DER POL derived a particularly convenient, frequency domain representation for the Hertz potential. The result may be written as

$$\Pi(s) = \frac{I(s)dl}{2\pi\varepsilon_0 s K\varrho}\frac{K^{1/2}}{K^2-1}\int_{v_1}^{v_2}\frac{\exp[-s\varrho v/(c_0 v_1)]}{(v^2-1)^{3/2}}v\,dv, \tag{5.57}$$

where

$$v_1 = ((K+1)/K)^{1/2}, \tag{5.58}$$

$$v_2 = (K+1)^{1/2}. \tag{5.59}$$

To obtain the impulse response, VAN DER POL integrated (5.57) by parts and noted that the inverse Laplace transform is then obvious. As an alternate procedure, (5.57) may be written as the difference of two integrals that each range to $+\infty$; then a change of variable yields

$$\begin{aligned}
\Pi(s) = {} & \frac{I(s)dl}{2\pi\varepsilon_0 s(K-1)\varrho}\int_1^\infty\frac{\exp(-s\varrho v/c_0)}{[v^2(K+1)-K]^{3/2}}v\,dv \\
& -K^{-1/2}\int_1^\infty\frac{\exp(-s\varrho vK^{1/2}/c_0)}{[v^2(K+1)-1]^{3/2}}v\,dv.
\end{aligned} \tag{5.60}$$

When K is a constant (i.e., the dielectric is lossless), the s-dependence of (5.60) is elementary. Time domain results may be obtained by interchanging the order of the v-integrals and an inverse Laplace transform. With impulse or step function excitations, both the Laplace transforms and the (time domain) v-integrals may be evaluated analytically. A step function current moment has the transform

$$I(s)dl = I_0 dl/s. \tag{5.61}$$

With this excitation substituted into (5.60), the following time domain result is obtained

$$\Pi(T) = 0, \qquad\qquad\qquad\qquad\qquad\qquad T < 1$$

$$\Pi(T) = \frac{\eta_0 I_0 dl}{2\pi(K^2-1)}$$

$$\times \left\{ T - 1 - (K+1)^{-1/2} \ln \left(\frac{T + [T^2 - K/(K+1)]^{1/2}}{1 + (K+1)^{-1/2}} \right) \right\}$$

$$1 < T < K^{1/2}$$

$$\Pi(T) = \frac{\eta_0 I_0 dl}{2\pi(K^2-1)} \left\{ T(K-1)/K - (K^{1/2}-1)K^{-1/2} \right.$$

$$\left. - (K+1)^{-1/2} \ln \left(\frac{(K^2+K)^{1/2}+K}{1+(K+1)^{1/2}} \right) \right\} \qquad K^{1/2} < T \quad (5.62)$$

where $T = c_0 t/\varrho$ and $\eta_0 = (\mu_0/\varepsilon_0)^{1/2}$. The corresponding vertical electric field component is given by

$$E_1(T) = 0, \qquad\qquad\qquad\qquad\qquad T < 1$$

$$= \frac{-\eta_0 I_0 dl}{2\pi(K^2-1)\varrho^2} T\{1 + K[T^2(K+1) - K]^{-3/2}\},$$

$$1 < T < K^{1/2}$$

$$= \frac{-\eta_0 I_0 dl}{2\pi K(K+1)\varrho^2} T, \qquad\qquad K^{1/2} < T. \quad (5.63)$$

A step-function current moment produces a linearly growing charge dipole, and at long times the linearly growing field is apparent in (5.63). Although a solution for the decaying step function may be obtained by convolution, it is not particularly relevant. By the time the source decay begins, the effects of finite conductivity become important, and the lossless dielectric model is no longer useful. It is worthwhile to note that the initial step discontinuity predicted by (5.63) can also be derived by inverting the lateral wave representation $E_{\text{lat}}(s)$ in (5.54). Therefore, the initial step is independent of conductivity.

To examine further the effect of the interface on the ground wave, Fig. 5.9 shows the pulse responses in three dielectric models A, C, and F. In this case the lateral separation is $\varrho = 1000\,\text{m}$, and the source and observer are immediately below the surface, $h = z = 0$. The plots on linear scales in Fig. 5.9a include the early time approximation E_1. The relative effects of the three dispersive models are more evident in the log scale plots of Fig. 5.9b. At low frequencies model C is

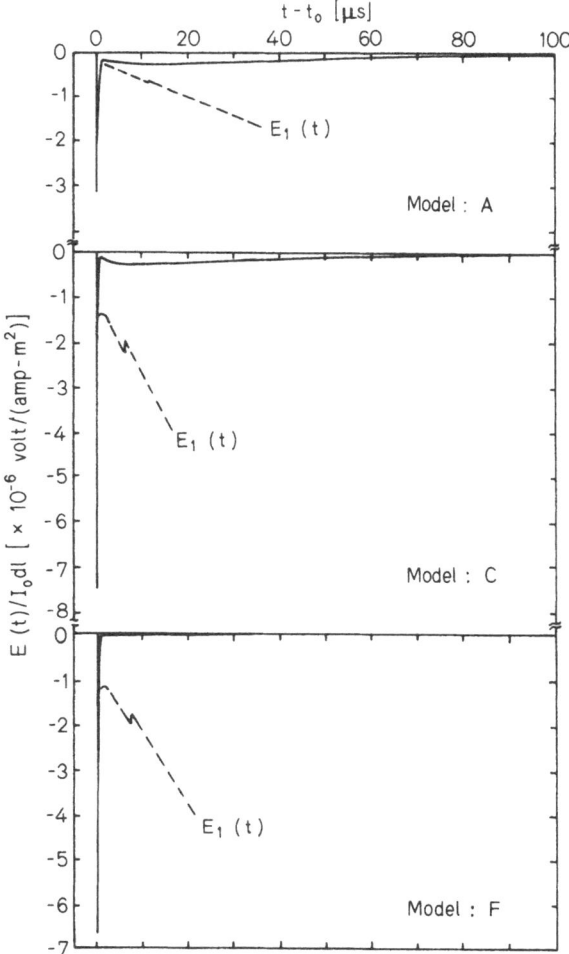

Fig. 5.9a. Pulse response in dispersive half-space media ($\varrho = 1000$ m; $h = z = 0$; $t_0 = 3.336$ μs)

equivalent to the "non-dispersive" model A, and at late times the equivalent pulse responses are evident. That model F is particularly dispersive at low frequencies is reflected by the long, slow tail of the pulse response. At high frequencies the dispersive models have conductivities of about 10^{-3} mho/m, while that of model A is only 10^{-5} mho/m. This explains the larger response of model A in the microsecond time range. The initial step discontinuities are indicated on the ordinate. Because the

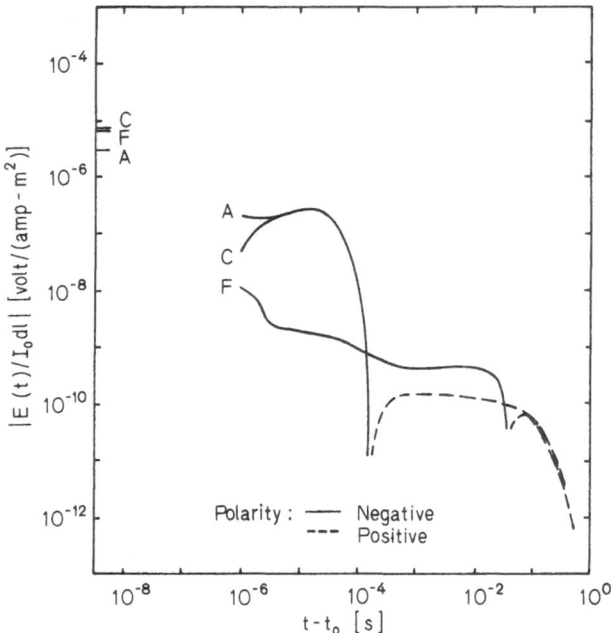

Fig. 5.9b. Pulse response in dispersive half-space media ($\varrho = 1000$ m; $h = z = 0$; $t_0 = 3.336$ μs)

antennas are at the surface, the path of the lateral wave is entirely through the air; in this case the early step is independent of the earth's conductivity. However, (5.63) shows that the step does vary inversely with $\varepsilon_e(\infty)/\varepsilon_0 - 1$, as the figures indicate. The plots in Fig. 5.9a are reproduced separately in Figs. 5.10, 5.11, and 5.12, where the approximations $E_0(t)$ and $E_1(t)$ are included. In dielectric A the pulse response agrees well with $E_1(t)$ until several microseconds have elapsed, but for C and F dispersion causes the responses $E(t)$ and $E_1(t)$ to diverge at earlier times.

Figure 5.13 shows the pulse response when the transmitter is buried at 100 m and the receiver is located just beneath the surface; the lateral separation is 100 m. This response was computed using the interface correction integral $P(D)$ in (5.41). At high frequencies excessive oscillation of the integrand introduced errors in the numerical computation of $P(D)$. As a consequence, the Laplace inversion produced a spurious oscillation during the first ten microseconds of the pulse. The oscillations are not shown in the figures.

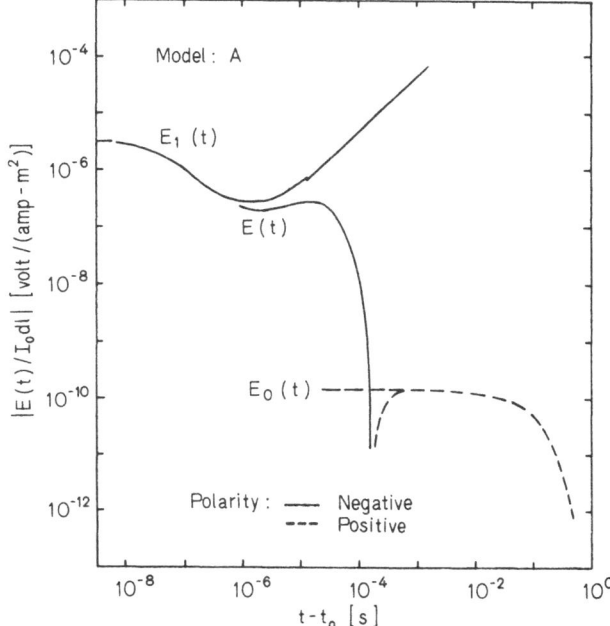

Fig. 5.10. Pulse response in a dispersive half-space, with approximations E_0 and E_1 ($\varrho = 1000$ m; $h = z = 0$; $t_0 = 3.336$ μs)

5.4 Discussion

Although a number of approximate solutions have been developed to describe pulse propagation in the earth, these generally fail to predict the effects of dispersive constitutive parameters. Restrictive approximations may be avoided by the application of careful numerical techniques to exact formulas.

Section 5.2 compares dispersive effects on pulse propagation between deeply buried antennas. Dispersion typical of dry rock or permafrost is shown to attenuate and delay the leading edge of the pulse waveform. Over a moderately dispersive 100 m path these effects occur during the first one-tenth microsecond after the pulse arrives, and over a 1000 m path they occur during the first microsecond. More severe, IP-type dispersion can have effects that extend into the millisecond time range.

Section 5.3 compares dispersive effects on propagation between antennas that are buried at shallow depths. A lateral wave, which travels

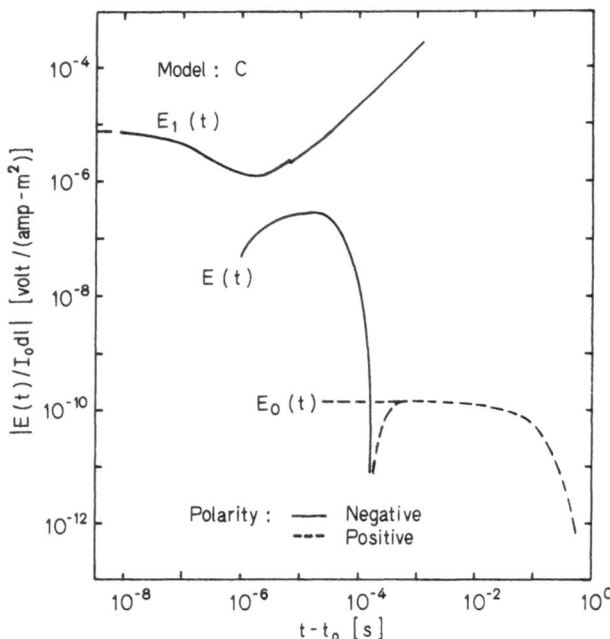

Fig. 5.11. Pulse response in a dispersive half-space, with approximations E_1 and E_0 ($\varrho = 1000$ m; $h = z = 0$; $t_0 = 3.336$ μs)

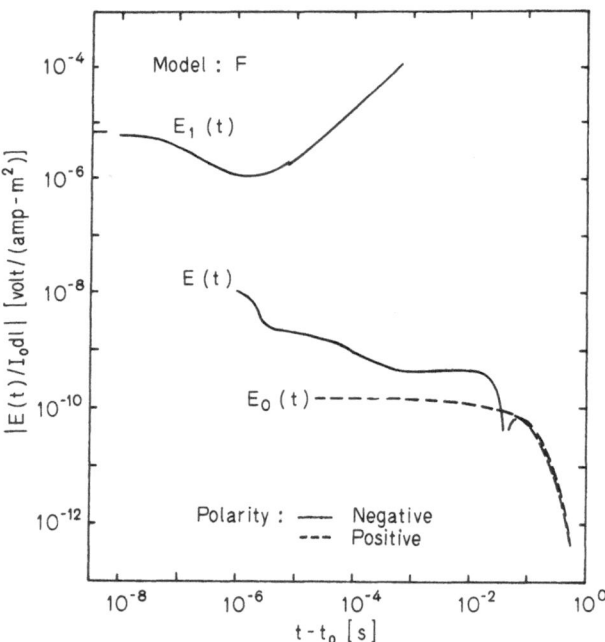

Fig. 5.12. Pulse response in a dispersive half-space, with approximations E_0 and E_1 ($\varrho = 1000$ m; $h = z = 0$; $t_0 = 3.336$ μs)

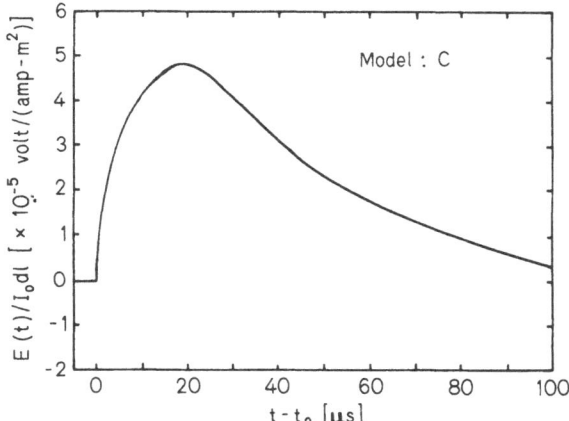

Fig. 5.13a. Pulse response in a dispersive half-space ($\varrho = 100$ m; $h = 100$ m; $z = 0$; $t_0 = 1.277$ µs)

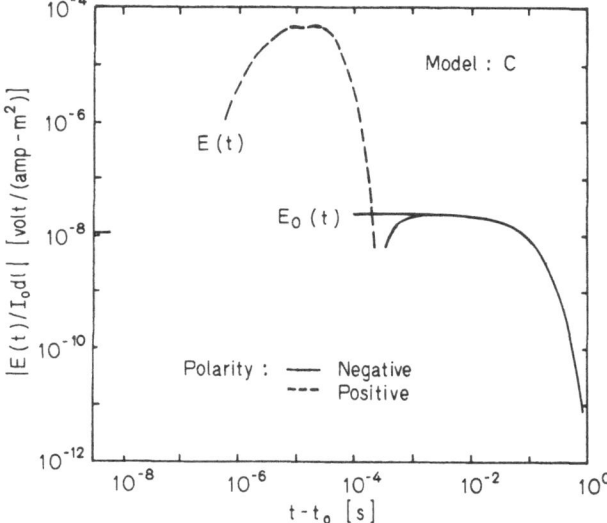

Fig. 5.13b. Pulse response in a dispersive half-space, with approximation E_0 ($\varrho = 100$ m; $h = 100$ m; $z = 0$; $t_0 = 1.277$ µs)

through the air, produces a rather large spike on the leading edge of the pulse waveform. At very long times the static field of a constant current element is observed. In dry rock dispersive effects are evident mainly during the lateral wave period, but with IP present, the effects extend into later phases of the pulse.

References

5.1   J. R. Wait: Electromagnetic fields of sources in lossy media, in *Antenna Theory*, Pt. 2, ed. by R. E. Collin and F. J. Zucker (McGraw-Hill, New York 1969) Chapt. 24

5.2   G. V. Keller: Electrical characteristics of the earth's crust, in *Electromagnetic Probing in Geophysics*, ed. by J. R. Wait (Golem Press, Boulder, Colo. 1971) Chapt. 1

5.3   J. H. Scott, R. D. Carroll, D. R. Cunningham: J. Geophys. Res. **72**, 5101 (1967)

5.4   R. N. Grubb, J. R. Wait: Electron. Lett. **7**, 506 (1971)

5.5   R. J. Lytle: "The Yosemite Experiments: HF Propagation Measurements Through Rock", Lawrence Livermore Lab., Report to AEC, no. UCRL-51381 (1973)

5.6   R. N. Grubb: Private communication (1972)

5.7   R. J. Lytle, E. F. Laine, D. L. Lager, J. Okada: "The Lisbourne Experiments: Propagation of HF Radio Waves through Permafrost Rock", Lawrence Livermore Lab., Report to AEC, no. UCRL-51474 (1973)

5.8   M. Saint-Amant, D. W. Strangway: Geophysics **35**, 624 (1970)

5.9   D. H. Chung, W. B. Westphal, G. Simmons: J. Geophys. Res. **75**, 6524 (1970)

5.10 G. V. Keller, P. H. Licastro: U. S. Geol. Survey Bull. **1052-H** 257 (1959)

5.11 J. R. MacDonald: J. Chem. Phys. **54**, 2026 (1971)

5.12 R. Alvarez: Geophys. **38**, 920 (1973)

5.13 J. R. Wait (Ed.): *Overvoltage Research and Geophysical Applications* (Pergamon Press, New York 1959)

5.14 T. R. Madden, D. J. Marshall: "Induced Polarization; a Study of Its Causes and Magnitudes in Geologic Materials", MIT Report to AEC, no. RME-3169 (1959)

5.15 L. A. Anderson, G. V. Keller: Geophysics **29**, 848 (1964)

5.16 J. R. Wait: A phenomenological theory of overvoltage for metallic particles, in *Overvoltage Research and Geophysical Applications*, ed. by J. R. Wait (Pergamon Press, New York 1959) Chapt. 3

5.17 R. E. Collin: *Field Theory of Guided Waves* (Mc-Graw-Hill, New York 1960) Chapt. 12

5.18 J. R. Wait: Geophysics **23**, 144 (1958)

5.19 A. J. Dekker: *Solid State Physics* (Prentic-Hall, Englewood Cliffs, N. J. 1957) Chapt. 6

5.20 B. D. Fuller, S. H. Ward: IEEE Trans. GE-**8**, 7 (1970)

5.21 C. A. Diaz: J. Geophys. Res. **77**, 4945 (1972)

5.22 R. T. Shuey, M. Johnson: Geophysics **38**, 37 (1973)

5.23 T. J. Katsube, R. H. Ahrens, L. S. Collett: Geophysics **38**, 106 (1973)

5.24 T. R. Madden, T. Cantwell: Induced polarization, a review, in *Mining Geophysics*, Vol. 2: Theory (Soc. of Expolaration Geophysicists, Tulsa 1967) pp. 373–400

5.25 D. F. Tuttle, Jr.: *Network Synthesis*, Vol. 1 (John Wiley and Sons, New York 1958) Chapt. 8, pp. 368–502

5.26 J. R. Wait: Radio Sci. **5**, 1461 (1970)

5.27 W. R. Eberle: "The effects of water content and water resistivity on the dispersion of resistivity and dielectric constant in quartz sand in the frequency range 10^2–10^8 Hz", *EMP Theoretical Notes*, Note 82, Air Force Weapons Lab., Kirtland AFB, N. M. (1971)

5.28 V. V. Daniel: *Dielectric Relaxation* (Academic Press, London 1967)

5.29 B. K. Bhattacharyya: Geophysics **22**, 905 (1957)

5.30 C. R. Burrows: IEEE Trans. AP-**11**, 286 (1963)

5.31 J. R. Wait, K. P. Spies: Canad. J. Geophys. **48**, 1858 (1970)

5.32 Y. L. LUKE: Proc. Cambridge Phil. Soc. **50**, 269 (1954)
5.33 H. L. GRAY, T. A. ATCHINSON: SIAM J. Num. Analysis **4**, 363 (1967)
5.34 J. R. WAIT, J. A. FULLER: IEEE Trans. AP-**19**, 796 (1971)
5.35 A. SOMMERFELD: *Partial Differential Equations in Physics* (Academic Press, New York 1949) Chapt. 4
5.36 M. ABRAMOWITZ, I. A. STEGUN (eds.): *Handbook of Mathematical Functions* (Appl. Math., Ser. 55) (NBS, Washington, D. C. 1964) Chapt. 9
5.37 G. N. WATSON: *A Treatise on the Theory of Bessel Functions*, 2nd ed. (Cambridge University Press, Cambridge 1966)
5.38 T. S. SHAO, T. C. CHEN, R. M. FRANK: Mathematics of Computation **18**, 598 (1964)
5.39 B. L. VAN DER WAERDEN: Appl. Sci. Res. B (Netherlands) **2**, 33 (1951)
5.40 A. BANOS, Jr.: *Dipole Radiation in the Presence of a Conducting Half-Space* (Pergamon Press, Oxford 1966) Sect. 3.2
5.41 J. A. FULLER: Electromagnetic Radiation and Propagation in the Presence of Dispersive Geological Media, Ph. D. dissertation, University of Colorado, Boulder (1974). (Available from University Microfilms, Ann Arbor, Mich.)
5.42 L. B. FELSEN: "Lateral waves" in *Electromagnetic Wave Theory*, Pt. 1, ed. by J. BROWN (Pergamon Press, London 1967) Chapt. A2, pp. 11–44
5.43 B. VAN DER POL: IRE Trans AP-**4**, 288 (1956)

Subject Index

Applied Physics

A monthly journal

Board of Editors	**A. Benninghoven,** Münster · **R. Gomer,** Chicago, Ill.
	H. K. V. Lotsch, Heidelberg · **H. J. Queisser,** Stuttgart
	F. P. Schäfer, Göttingen · **A. Seeger,** Stuttgart
	K. Shimoda, Tokyo · **T. Tamir,** Brooklyn, N.Y.
	H. P. J. Wijn, Eindhoven · **W. T. Welford,** London

Coverage

application-oriented experimental and theoretical physics:

Solid-State Physics	*Quantum Electronics*
Surface Physics	*Laser Spectroscopy*
Infrared Physics	*Photophysical Chemistry*
Microwave Acoustics	*Optical Physics*
Electrophysics	*Integrated Optics*

Special Features

rapid publication (3-4 months)
no page charges for **concise** reports

Languages

Mostly English

Articles

review and/or tutorial papers
original reports, and short communications
abstracts of forthcoming papers

Manuscripts

to Springer-Verlag (Attn. H. Lotsch), P.O. Box 105 280
D-69 Heidelberg 1, F.R. Germany

Place North-American orders with:
Springer-Verlag New York Inc., 175 Fifth Avenue, New York. N.Y. 10010, USA

Springer-Verlag
Berlin Heidelberg New York

Topics in Applied Physics

Founded by H. K. V. Lotsch

Volume 3:

Numerical and Asymptotic Techniques in Electromagnetics

Edited by R. Mittra

Contents: R. Mittra: Introduction. – W. A. Imbriale: Applications of the Method of Moments to Thin-Wire Elements and Arrays. – R. F. Harrington: Characteristic Modes for Antennas and Scatterers. – E. K. Miller, F. J. Deadrick: Some Computational Aspects of Thin-Wire Modeling. – R. Mittra, C. A. Klein: Stability and Convergence of Moment Method Solutions. – R. G. Kouyoumjian: The Geometrical Theory of Diffraction and Its Application. – W. V. T. Rusch: Reflector Antennas. – Subject Index.

K. D. Becker

Ausbreitung elektromagnetischer Wellen

Eine Einführung in die Theorie
Hochschultext. Edited by W. Rupprecht.

The properties of the propagation of electromagnetic waves are examined on the basis of Maxwell's equations and the fundamentals of the theory of antennae are set out. The propagation of ultrashort waves in homogeneous and inhomogeneous atmospheres is discussed. The propagation of long waves and the influence of the ionosphere are also dealt with, as is the question of whether surface waves exist and, if so, what properties they possess.

Springer-Verlag Berlin Heidelberg New York